HEALTH SYSTEMS RESEARCH

Edited by K. Davis and W. van Eimeren

W. van Eimeren B. Horisberger (Eds.)

Socioeconomic Evaluation of Drug Therapy

With 52 Figures

Springer-Verlag Berlin Heidelberg GmbH

Prof. Dr. Wilhelm van Eimeren
MEDIS
Ingolstädter Landstraße 1

D-8042 Neuherberg

Dr. Bruno Horisberger
Interdisciplinary Research Centre
for Public Health
Rorschacher Straße 103 C

CH-9007 St. Gallen

ISBN 978-3-642-64811-3   ISBN 978-3-642-61366-1 (eBook)
DOI 10.1007/978-3-642-61366-1

Preface

In past decades little attention was paid to the social and economic dimensions of health care. Health as a basic human need was mainly assessed from a clinical point of view. In addition, the dominant assumption was that the demand for health services is limited and calculable. The conclusion was drawn that the necessary financial resources would be both available and forthcoming.

Reality has shown this conclusion to be illusory. Health expenditure in all countries has roughly doubled as a proportion of national wealth in the past 25 years, and the proportion may double again in the next 25. By the year 2010 U.S. citizens may on average be spending one dollar out of every five on health care. Consequently there is worldwide concern over the best and most appropriate ways in which resources should be allocated in the health care sector. This concern has led decision-makers and opinion leaders to consider ways of subjecting health care practices to closer social and economic scrutiny. For pharmaceutical preparations, for example, it is no longer sufficient to demonstrate quality, efficacy, and safety. Increasingly attention is focused on additional parameters: Notably, evidence of cost-benefit and quality of life. In an era of rapidly rising health care costs, choices are inevitably going to be made on therapeutic as well as on economic grounds. In addition, however, it will be mandatory to show that the benefits of a medicine, in terms of savings in other health care resources or in improved health, justify the costs, and that the product fulfills the basic efficiency requirement of "value for money."

To discuss these trends and latest developments in the socioeconomic evaluation of drug therapy, the Interdisciplinary Research Centre for Public Health, St. Gallen, and HealthEcon Ltd., Basel, in a joint effort organized a symposium in Wolfsberg, on the shores of Lake Constance. This symposium strengthens the tradition of Wolfsberg meetings dealing with issues of health care, e.g., cost sharing and technology assessment. An international group of experts in medicine, economics, epidemiology, and sociology were invited to present actual results of their studies and open them to interdisciplinary exchange of views. The meeting, whose proceedings are recorded in this book, provided a unique platform for the discussion of the background, the potential, and the limitations of socioeconomic appraisal of drug therapy. The papers presented are divided into four parts.

Part I gives an introduction to the role of economics in drug therapy. Some papers profile the changing landscape of economy and medicine as it affects the economies of health and pharmaceutical industries. They focus on how care is organized and paid for, especially in regard to pharmacotherapy. Examples illustrate how cost-effectiveness and quality of life considerations influence decisions about coverage, provision, and reimbursement for particular agents and products. Other papers cover basic issues such as the pricing policy of the pharmaceutical industry and the social and economic implications of diseases. A concluding paper looks at some of the earlier ways in which drug therapy has been assessed and emphasizes the insensitivity and limited focus of traditional end points in regard to economic and social outcomes.

Socioeconomic evaluation of drug therapy cannot be carried out without attention to practical considerations. For the next two sessions a drug therapy was selected that had been subjected to extensive and differentiated evaluation: the transdermal therapeutic system Nitroderm TTS. This patch was introduced to the market in 1982, placing at the disposal of doctors for the first time a preparation that enables nitroglycerin to be administered through the skin in a controlled dosage for the long-term prophylaxis of angina pectoris attacks. Because the costs per unit and per dosage of this system are higher than those for the traditional nitrate preparations in tablet, capsule, or ointment form, evidence had to be provided substantiating the product's "value for money." Today, Nitroderm TTS is, from the socioeconomic viewpoint, one of the most thoroughly analyzed innovations in drug therapy.

Part II deals predominantly with the economic criteria involved in measuring the effectiveness of drug therapy. The first paper discusses the basic methodology and procedure of economic drug appraisal, pointing out the difficulties in employing an ideal study design. After a comprehensive overview of the studies to follow, several papers focus on description of single studies, their methodologies, and their results.

"If you want to know something about the patient's disease, you'd better stop asking the doctor, you'd better ask the patient." This statement opened the discussion at the symposium on the social and qualitative criteria involved in measuring the effectiveness of drugs, reported in Part III. Again, there is a paper outlining the basic approaches to measuring quality of life as well as other nonmonetary considerations, such as compliance and patient preference. Several papers which follow deal with empirical studies in different settings relevant to these subjects.

Each paper in the plenary session was followed by a discussion. In addition, after Parts II and III, a discussant summarized the key features of the different studies and tried to unify them. Also, discrepancies and methodological shortcomings were pointed out and placed in perspective. Of course, there is insufficient space for extended reporting of these contributions and personal interventions. The editors, instead, have sought to draw out the essence of the discussions in producing brief résumés placed at the end of Parts I, II, and III.

In Part IV the outlook for socioeconomic evaluation of drug therapy is discussed. The scientific, the societal, and the industrial points of view are presented.

Part V concludes the book with a general essay by the organizers of the symposium and with the curricula vitae of the authors. Part V also contains a very extensive index section for those who either seek specific information or want to follow a topic that the book itself does not use as a structural principle.

Acknowledgments

First of all we are obliged to the participants, whose papers and lively discussions made the symposium a success. Particular thanks go to the joint moderators of the three half-day sessions: Dr. B. Horisberger and Dr. R. Dinkel (part I), Dr. L. Read and Dr. B. Luce (part II), and Prof. S. Walker and Prof. C. R. B. Joyce (part III). Special appreciation is due to Susan Inderbinen, Gaby Lederer, and Alfred Geiser, who were responsible for all symposium logistics and secretarial work.

We owe a great debt to Ciba-Geigy, who supported the conference, and also to Prof. W. von Wartburg, who initiated it. Finally we owe a debt of gratitude to the European Office of the World Health Organization, for support and technical assistance in preparing and holding the symposium, and to Dr. Kaprio, special adviser to WHO Headquarters in Geneva, Switzerland, for his most valuable contribution.

Contents

Part I: The Role of Economics in Drug Therapy

Part II: Measuring the Effectiveness of Drug Therapy –
The Economic Criteria

**Part III: Measuring the Effectiveness of Drug Therapy –
Social and Qualitative Criteria**

**Part IV: Panel Discussion on the Outlook
for Socioeconomic Evaluation of Drug Therapy**

Part V: Envoi

List of Participants

Prof. FRITZ BESKE
  Institute for Health Systems Research, Beselerallee 41, D-2300 Kiel 1,
  Federal Republic of Germany

Dr. BERNARD S. BLOOM
  Research Associate Professor, University of Pennsylvania, Leonhard Davis Institute of Health Economics, Colonial Penn Centre, Philadelphia, PA 19104, USA

PETER F. CARPENTER
  President and Director Alza Development Corporation, 950 Page Mill Road,
  P. O. Box 10950, Palo Alto, California 94303–0802, USA

Dr. ROLF DINKEL
  Member of Management Committee, HealthEcon Ltd., Steinentorstraße 19,
  CH-4051 Basel, Switzerland

Prof. WILHELM VAN EIMEREN
  MEDIS, Ingolstädter Landstraße 1, D-8042 Neuherberg,
  Federal Republic of Germany

Prof. HANS FRIEBEL
  Temporary Adviser to WHO Regional Office for Europe, Copenhagen,
  Uferstraße 49, D-6900 Heidelberg, Federal Republic of Germany

Dr. TONI GRAF-BAUMANN
  Editor, Springer-Verlag, Tiergartenstraße 17, D-6900 Heidelberg 1,
  Federal Republic of Germany

Dr. B. GÜTHER
  Infratest Health Research, Landsberger Straße 338, D-8000 München 21,
  Federal Republic of Germany

ROBERT HANKIN
  Commission of the European Communities, Directorate-General III/B/3,
  Rue de la Loi 200, B-1049 Brussels, Belgium

Dr. BRUNO HORISBERGER
  Interdisciplinary Research Centre for Public Health, Rorschacher Str. 103 C,
  CH-9007 St. Gallen, Switzerland

Prof. BENGT JÖNSSON
  Linköping University, Centre for Medical Technology Assessment,
  S-581 83 Linköping, Sweden

Prof. C. R. B. JOYCE
  Head of Project Innovation, Ciba-Geigy Ltd., CH-4002 Basel, Switzerland

Dr. LEO KAPRIO
  Special Adviser to the Director General, Headquarters,
  World Health Organization, Av. Appia 20, CH-1211 Geneva 27, Switzerland

Prof. BENNO KÖNIG
  Johann-Gutenberg-University Mainz, Am Pulverturm 13, D-6500 Mainz 1,
  Federal Republic of Germany

Dr. PETER LAUPER
  Head of Pharma Economics, Ciba-Geigy Ltd., CH-4002 Basel, Switzerland

DAVID LOSHAK
  Medical Writer, 169 Half Moon Lane, London SE24 9JG, Great Britain

Dr. BRYAN R. LUCE
  Senior Research Scientist, Battelle, Human Affairs Research Centers,
  2030 M Street N. W., Washington, D. C. 20036-3391, USA

Dr. MARIE E. MICHNICH
  Associate Executive Vice President, American College of Cardiology,
  Old Georgetown Road, Bethesda, Maryland 20814, USA

Prof. DUNCAN NEUHAUSER
  Department of Epidemiology and Biostatistics, School of Medicine,
  Case Western Reserve University, Cleveland, Ohio 44106, USA

Prof. PETER O. OBERENDER
  University Bayreuth, Chair of Economics, D-8580 Bayreuth,
  Federal Republic of Germany

Dr. GERRY OSTER
  Policy Analysis Inc., 1577 Beacon Street, Brookline, MA 02146, USA

Dr. J. LEIGHTON READ
  New England Deaconess Hospital, 194 Pilgrim Road, Boston,
  Massachusetts 02215, USA

Prof. BEAT ROOS
  Director, Swiss Federal Office of Public Health, Bollwerk 27,
  CH-3001 Bern, Switzerland

Dr. HEINZ SCHNEIDER
  Project Manager, HealthEcon Ltd., Steinentorstraße 19, CH-4051 Basel,
  Switzerland

Dr. DAVID TAYLOR
  Director of Public and Economic Affairs, Association of the British Pharmaceutical
  Industry (ABPI), 12 Whitehall, London SW1 2 DY, Great Britain

Dr. STANLEY H. TAYLOR
The General Infirmary at Leeds, Great George Street,
Leeds LS1 3EX, Great Britain

Prof. W. P. VON WARTBURG
Member of the Executive Committee, Ciba-Geigy Ltd.,
CH-4002 Basel, Switzerland

Prof. STUART R. WALKER
Director, Centre for Medicines Research, Woodmansterne Road,
Carshalton, Surrey SM5 4DS, Great Britain

Prof. ELLEN WEBER
Director of Clinical Pharmacology of Ludolf Krehl
Clinic of the University of Heidelberg, Bergheimerstraße 56a,
D-6900 Heidelberg, Federal Republic of Germany

Dr. GAIL R. WILENSKY
Center of Health Affairs, Project HOPE, 2 Wisconsin Circle, Suite 500,
Chevy Chase, MD 20815, USA

Prof. PETER ZWEIFEL
Institute for Empirical Research in Economics, University of Zürich,
Kleinstraße 15, CH-8008 Zürich, Switzerland

Opening Adress

B. Roos

I have come here from Bern, a wonderful trip on this spring morning, and I am very pleased to welcome you to Wolfsberg on behalf of the Federal Government, especially the Swiss Federal Department of the Interior and its head, Federal Councillor Flavio Cotti. As you know, Switzerland is a federal state with 26 counties or cantons, all having competence in public health matters. At the federal level we are "only" concerned with health protection, which is mainly based on federal legislation.

I am pleased to greet all those present and especially the great number of foreign guests who are here to discuss questions which cause deep concern not only in Switzerland but internationally. Credit and thanks are primarily due to the Regional Office for Europe of the WHO for their support and technical assistance in preparing and holding the symposium. I am pleased to welcome my old friend – I now call him an elder statesman – Dr. Leo Kaprio, the former Director of the WHO Regional Office for Europe. After Dr. Kaprio retired from his office in Copenhagen he actually became the man behind the scenes in the drug strategy of the WHO. He is very familiar with the problems addressed by the World Health Assembly 4 years ago in a resolution on "Rational Use of Drugs," and he is familiar with all the meetings and discussions held since then. It is to the merit of Dr. Kaprio that all interested parties can now discuss these important problems which face developing countries as well as industrialized countries, while restraining themselves from ideological statements. Thank you very much, Dr. Kaprio, for coming. And then I would like to welcome Professor Friebel from Germany. He is a temporary advisor to the WHO in the same field. And of course, I would also like to thank the promoters and organizers of this meeting, especially the Interdisciplinary Research Centre for Public Health in St. Gallen headed by my friend, Dr. Bruno Horisberger, and HealthEcon Ltd., Health Service Consultants in Basel.

They have chosen a stimulating venue, attracting an international and interdisciplinary group of experts in medicine, economics, epidemiology, and sociology. Twenty years ago we said naively "more must be better" in the provision of medical care. I think few people still believe in such a simple philosophy today. All countries need to make tough choices in how to use their scarce resources by asking whether more resources should be devoted to health care or, for instance, to education. I regard this as a major issue in the discussion of the internal politics of a developing country: Does education take priority over health care or vice versa? As you know, competition for resources occurs not only between health care and other sections of the economy but also within the health care sector itself. Should more resources be devoted to primary health care? Primary health care is a vehicle for achieving the goal of the WHO: Health for All by the Year 2000. A lot of discussion is also going on in our country on the merits of primary health care – in German we call it "Spitex": *Spitalexterne Betreuung* – as against the continuous promotion of high technology medicine in sophisticated hospitals. In an industrialized country it is easy for people to speak of ideology and primary health care. But as soon as they have a personal health problem, ideology is forgotten and they go to the best fitted hospital with all the equipment they need.

Within the treatment options for a given therapeutic indication should resources be allocated only to already available preparations or be expanded to innovative approaches? On this question I would like to make just a few remarks: The biggest challenge at the end of this century as Dr. Mahler, the Director General of the WHO, said, is the new disease AIDS. We have seen in one lifetime of a medical doctor the disappearance of poliomyelitis, due to the introduction of the Salk and Sabin vaccine, and now the appearance of a new disease. As you know, we have no therapeutic methods at hand, we have no vaccine. At the moment we discuss and we promote prevention, especially „mechanical" prevention by means of condoms. AIDS is a real challenge to all participants in the arena of health care.

Coming back to the topics and the questions of this symposium: None of us involved in health policy is going to pretend that choices about the needs of the patient, or about the costs, are easy to make. The difficulties in allocating limited resources between competing programs, projects, and interventions have led government and other partners in the health care sector to consider ways of subjecting health care practices to closer social and economic scrutiny. Nowadays, in many cases it is no longer sufficient to demonstrate that a new diagnostic or therapeutic approach has a positive clinical effect. It is also necessary to show that the benefits in terms of improved health, better quality of life, or savings in other health care resources in some sense justify the costs.

As an example of such a decision I would like to mention Norway, where the government has decided to introduce in its drug policy a so-called need clause for a new medication. I think this is not possible in Switzerland because our constitution stipulates free trade and free enterprise. Our industries, especially the pharmaceutical, have shown a marked sense of responsibility. When they produce a new drug they take more and more into account, not only the clinical effect but also the social and economic aspects. It is becoming standard practice in health-related industries to rely upon universities, academia, or health services research centers for cost-benefit analysis – be it in respect of drugs, vaccines, or medical equipment. Certain industries have even established their own in-house analytical capacity for economic appraisal. This increasing interest is also reflected by a growing number of theoretical publications on the subject. However, at the same time, a fundamental concern is noticeable with respect to the real operational usefulness of socioeconomic analysis as a decision-making tool in the formulation of health policy. Health policy makers who are interested in a more rational approach in the pursuit of cost-effective care are often confused by the claims concerning the validity of studies and by the numerous underlying methodological controversies. Under these circumstances, it is particularly creditable that this symposium intends to review socioeconomic appraisal of drugs from a theoretical and a practical point of view. The symposium provides a very useful platform for the necessary dialogue between the disciplines and the discussion of the rationale underlying cost-benefit and similar analyses as well as their potential and limitations. It is hoped that this meeting will contribute toward our understanding of the applicability and usefulness of socioeconomic evaluation of drug therapy and our capability to act upon outcomes of relevant studies as informed customers. I think it is very good that at this meeting you will be able to discuss the whole range of these problems. I hope you will find solutions and/or answers. It is important for you, coming from universities and from pharmaceutical companies and health policy making institutions to get answers which can be used in the discussion with other organizations involved in these problems – for example, the International Consumers Organization or "the other side" – I wouldn't say the opposition. Some of the opponents also have their arguments. It is, therefore, good that in future the discussion can be held on a scientific basis. The WHO will play its role as a platform where all these different circles can meet. We should never forget that it is a huge task for a lot of countries in the world simply to have drugs at hand.

I hope that you have fruitful discussions during this important symposium.

Part I

The Role of Economics in Drug Therapy

1. The Changing Economic Milieu in the United States

D. Neuhauser

Dr. Florence Wilson and I have spent 15 years attempting to describe *Health Services in the United States* and we are currently working on the fifth edition of a book with this title. We are still far from satisfied.

We have set ourselves the relatively modest task of describing the component parts of medical care in approximately 350 pages. Using the analogy of medical education, this can be considered an anatomy of health care. It stops short of being a physiology of medical care delivery: that is, how the component parts relate to each other. Describing this physiology of medical care in the United States is my assignment for this paper. Fortunately, this falls short of being a pathology of medical care delivery: that is, the diseases and the failures of the medical care system. It also excludes clinical medicine i. e., how to cure or palliate these diseases.

Given 15 years of work, perhaps I could describe this physiology in 350 pages. This effort falls short of such an aspiration.

My physiological model contains 39 component parts (see Table 1.1) and 63 relationships (see Fig. 1.1, parts I and II). The choices of these parts and relationships are far from exhaustive. The list focuses on important changes currently in process. For this reason the parts often describe changes, rising costs, more doctors, more competition, rather than a static condition. One can imagine the appropriate mathematics as a spectacular number of simultaneous differential equations.

Unlike economic models that define a flow of money in and out of component parts, the parts define different phenomena: economic, organizational, behavioral, and even perceptions. All of these are needed to understand the current American scene.

The reader may look at what follows and decide that new parts and relationships should be added. However limited the model that follows, I believe I have successfully demonstrated the complexity of the changes under way.

So much for apologies and explanations. Now for the model. The component parts are divided into five groups. These are: the general environment [1–8], the health care environment [9–19], new medical care system responses [20–27], new provider responses [28–37], and effects on the public [38, 39].

The relationships between these parts are numbered from 101 to 163 and shown in Fig. 1.1, parts I and II. Some of these relationships have causal direction arrows and some have "+" or "−" signs to indicate the positive or negative association. A description of the model follows.

Fig. 1.1. A model of medical care delivery in the United States, part I

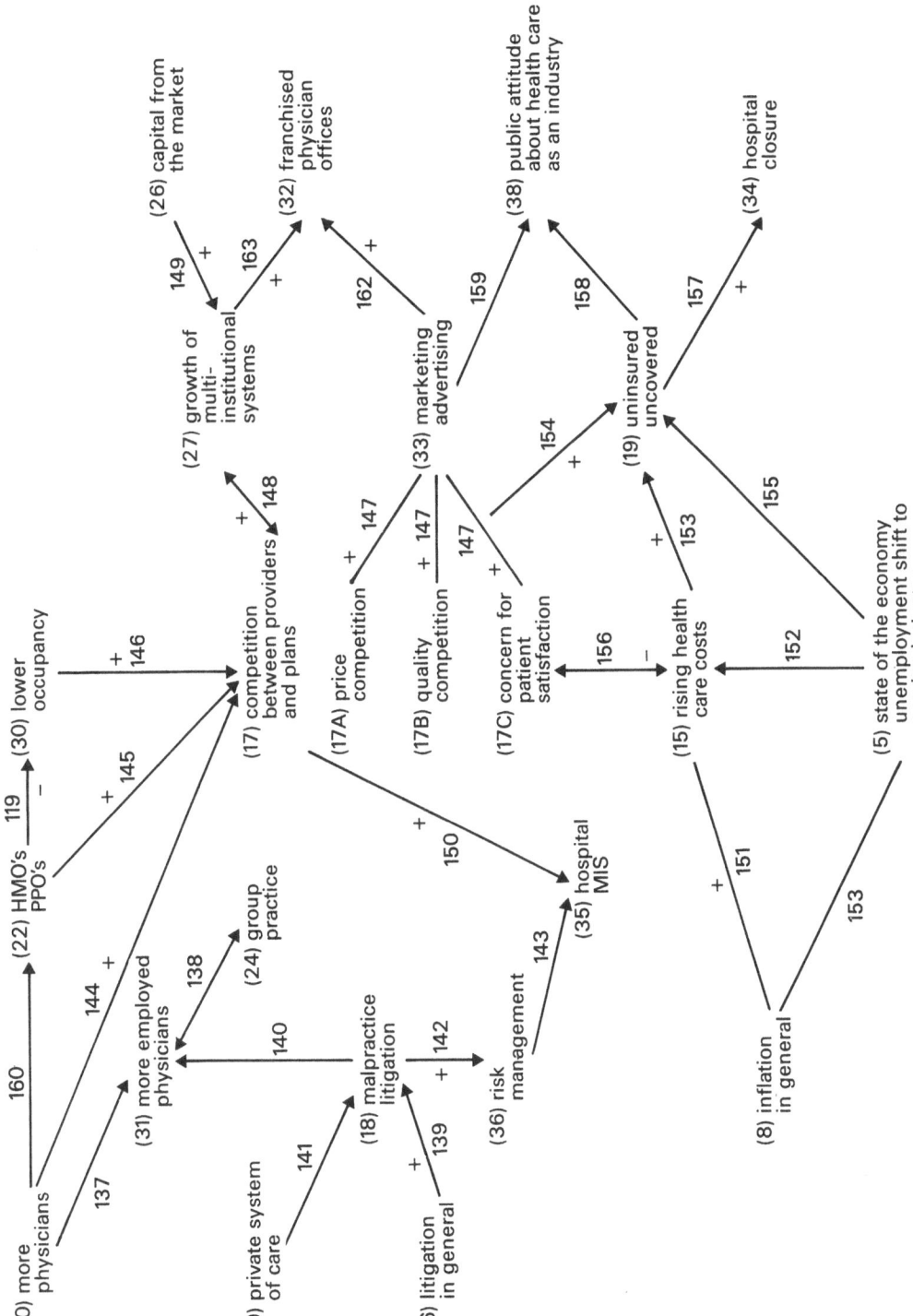

Fig. 1.1. A model of medical care delivery in the United States, part II

Table 1.1. Components of the medical care system in the United States

General environment (exogenous)

1. Population size and growth
2. Aging population
3. Changing disease, AIDS
4. Employer payment and employee choice
5. Changes in the economy
6. Litigation climate
7. General advance of technology
8. Inflation

Health care environment

9. Medical care delivered privately
10. Increase in number of doctors
11. Medical research
12. Manifest burden of illness
13. Costs and use of technology in medical care
14. Health insurance (private, Medicare, Medicaid)
15. Rising health care costs
16. Employer involvement in cost control
17. Competition between providers
    17A) Price competition
    17B) Quality competition
    17C) Concern for patient satisfaction
18. Medical malpractice litigation
19. Growth of uncompensated care, uninsured

New medical care system responses

20. Cost control measures
21. DRG payment
22. HMOs, PPOs
23. Ambulatory surgery
24. Group practice
25. Home care, hospice, DMS
26. Capital obtained from money market
27. Growth of multi-institution systems

New responses from providers of care

28. Shorter Length of stay
29. Fewer hospital admissions
30. Lower hospital occupancy
31. More physicians as employees
32. Franchised physician offices
33. More advertising, marketing
34. Hospital mergers and closures
35. Growth of hospital management information systems
36. Growth of risk management
37. Hospital profitability

Effects on the public

38. New perceptions of health care as a business
39. The health of the public

The American population stands at 247 million people and is growing ([1], 101). It is also an aging population ([2], 102). A major change in disease prevalence ([3], 103) is the growth of AIDS, although its present impact is small. This burden of illness [12] affects the total cost of technology and pharmaceuticals in use ([13], 136), and both in turn result in the rising costs of health care ([15], 107, 129). The advance of technology [7] (computers, lasers, super conductivity) in part drives medical research ([11], 104) and both determine the costs of technology ([13], 105, 130). This is not always the case. Smallpox vaccination has eliminated the cost of this disease worldwide ([15], 129).

It is central to understanding U.S. medical care to realize that care is largely provided by nongovernment, private organizations [9] such as hospitals, nursing homes, physician group practices, and health insurers. This is associated with a traditional separation of health insurance (third party) payment to independent providers (second party) for care of patients (first party) [14]. This separation of payment and care also characterizes the two major federal government payment programs: Medicare for the elderly and Medicaid for the poor (106).

Health insurance [14] during the past 50 years was created due to the development of costly technology ([13,15], 108, 109, 129) and makes development of further technology possible. The association of technology, health care costs, and government payment ([13–15], 108, 109, 129) has been the basis for the worldwide health care technology assessment movement ([20], 115).

The private delivery system [9] is associated with the choice of a health insurance plan through the worker's place of employment ([4], 110), with multiple competing providers offering to pay and/or provide health care ([17], 134). Because employers pay part or all of health care costs [4], many employers have become directly involved in shaping medical care delivery toward lowering the high costs of care ([4, 15, 16], 111, 112). This has included promotion of Health Maintenance Organizations (HMOs), Preferred Providers (PPOs) ([22], 133) and cost control mechanisms [16] such as preadmission review ([21], 131).

In the same way, insurers [14] and the government have promoted cost containment measures ([20], 114). One such notable example is payment by diagnostic-related groups or DRGs [21].

Health Maintenance Organizations combine payment and provision of care by private organizations [22]. Employees and their families who choose one pay fixed monthly premiums regardless of care received. The HMO provides their care but these members are not ordinarily reimbursed for care outside the HMO. One result of care by HMOs is reduced use of hospital care (119).

Preferred Providers differ in that they pay for care outside the nonpreferred providers, but with a large self-pay component. The PPO contracts with fee-for-service physicians and some hospitals. The patient may get full coverage for using them but must pay 20% or 30% of care provided by other nonpreferred providers. Preadmission review and DRG payment [21] by Medicare and others have sharply reduced length of patient stay in hospital ([28], 118).

Ambulatory surgery ([23, 116) can lower costs (128) and avoid hospital admissions ([29], 127). Shorter length of stay has led to a growth of the use of home care ([25], 125), including hospice care, particularly for terminally ill cancer patients. This has led to growth in sales of durable medical supplies and equipment that can be used at home ([13], 126). The DRG reimbursement system [21] compels hospitals to understand the costs of care for each case. One result is a new infusion of effort to upgrade hospital management information systems (MISs) ([35], 117).

Shorter stay [28] and fewer admissions [29] have resulted in lower hospital occupancy ([30], 120, 121). Fewer admissions and lower occupancy [30] have resulted in more closures and mergers of hospitals ([34], 122) with the expectation of more to come. In the short run, shorter length of stay and DRG payments have resulted in excess revenue for hospitals ([37], 123, 124), but this is not expected to last.

As in many other countries, an increased number of physicians [10] are being educated in the United States. This has led to and will continue to promote competition ([17], 144). Physicians will be more interested in working for alternative delivery systems like HMOs and PPOs ([22], 160). They may be more interested in employment in groups, ([24, 31], 137, 138), rather than in solo practice. The lower hospital occupancy resulting from HMOs ([22, 30], 119) has promoted more intensive competition ([17], 144) among hospitals.

The United States is a litigious society with lots of lawyers and lawsuits [6]. This has its impact on medical care through the growing costs of medical malpractice litigation ([18], 139). Because of our private system of care [9] and lack of extensive welfare benefits, plaintiffs sue for larger amounts of money to pay for the costs of further care or lost income not provided by the State ([18], 141). This makes the malpractice problem a larger one. The result is expensive malpractice insurance premiums for physicians. The new physician cannot afford these costs and may seek employment in a group practice as a result ([24, 31], 138). Another response to malpractice costs [18] is the development of risk management ([36], 142) as a range of activities related to law, quality control, staff education, and accident avoidance. This promotes more computer-generated management information ([35], 143). Quality competition [17B] is compelling hospitals to demonstrate the quality of care they provide, thereby increasing the demand for information (150).

Competition between providers [17] can be divided into three components: price competition [17A], quality competition [17B], and a new concern for patient satisfaction [17C].

Being able to compete on price has led to a concern for costs [15].

Rising costs have encouraged competition [17A]. Competition has created a concern for marketing and an explosive growth in medical care provider advertising ([17B], 147). Concern for patient satisfaction ([17C], 156) has resulted in attention to hospital environments and waiting and travel times.

Competition by private providers [17] means that providers are compelled to reduce care for patients who are not covered by health insurance ([19], 154). Competition [17] has promoted the growth of multi-institution systems ([27], 148) through both vertical and horizontal integration.

Because private providers must seek capital to expand and build, and because this capital comes from the money markets [26], those providers which can obtain money at lower interest rates and costs can compete more successfully. Multi-institution systems can obtain money at lower cost ([27], 149).

Although this trend has scarcely started, these systems [27] and the need for advertising [33] will probably promote the growth of franchised physician offices ([32], 162, 163) and emergicenters, sometimes referred to as "doc in a box."

Inflation in the general economy [8] and the state of the economy as a whole [5] influence the costs of medical care ([15], 151–153). The decline of manufacturing and the growth of service jobs [5] has increased the number of jobs that include extensive health insurance coverage. These changes have also led to unemployment.

These trends plus the high costs of care [15] have increased the number of uninsured people ([19], 153, 155). Hospitals in areas with higher proportions of uninsured people are at higher risk of closure ([34], 157).

More uninsured people [19] and the perception that medical care is a business [38] may result in shifts in public attitudes about medical care in a competitive environment [33]. This could lead to public pressure to move away from competition toward more government regulation (158, 159).

One characteristic of American medical care, which is not included in the diagrams, is the effect of medical care on the health of the public [39]. This deserves a special comment.

Two recent studies of the relation between medical care and the people's health in the United States are worthy of note: Jack Hadley's econometric study suggests that more medical resources put into the most underserved parts of the country would improve health. The Rand Insurance trial shows that middle-income families could use less medical care and their health would not be affected. However, the poor would benefit from more medical care.

The model we have described here can certainly be elaborated. One could emphasize some parts over others. Let me highlight a few parts with relevance to the socioeconomics of pharmaceuticals.

Pay careful attention to the needs of market-driven multiinstitution systems. This is not only a question of large volume purchasing and inventory levels, but also a clear understanding of the multis' market strategy. The multis are only beginning to develop system-wide bacteriology laboratories, franchised physician offices, and market segmentation, but these changes can be prepared for.

Any technology that shortens length of stay will be favorably considered. Products related to home care are likely to be in demand.

The manufacturer should be well aware of the needs of medical insurers to understand under what conditions their products should be covered by insurance. There is a lot of experimentation going on in insurance benefits, providing opportunity for new approaches.

A simple analysis of drug benefit is no longer adequate. Cost-effectiveness narrowly defined is also too limited. A new product that can shorten a hospital stay by a day can command a large premium in the market. The multi-institution systems interested in having happy, satisfied patients may not be interested in drugs that slightly prolong life at the price of unpleasant side-effects. The new American environment more than ever is calling for a broad-based socioeconomic evaluation of pharmaceuticals and technologies. Perhaps on that note, it is the right moment to turn to such broad-spectrum efforts to evaluate one new product in particular: transdermal nitrate. Can these efforts serve as an example of what needs to be done? Read on.

Bibliography

This bibliography is keyed to the numbered parts and relationships in Table 1.1 and Fig. 1.1. It is purposefully very selective.

Wilson F, Neuhauser D (1985) *Health services In the United States,* 2nd edn. Ballinger, Cambridge, Mass., pp 337

[1], [2] Fuchs VR (1983) *How we live.* Harvard University Press, Cambridge, Mass. U.S. Department of Health and Human Services, Public Health Service (1986) *Health United States 1986.* Hyattsville, Maryland
[4] Enthoven AC (1980) *Health plan.* Addison Wesley, Reading, Mass.
[11] Mushkin S (1979) *Biomedical research: costs, and benefits.* Ballinger, Cambridge, Mass.
[13] Reiser S, Anbar M (1984) *The machine at the bedside.* Cambridge University Press, Cambridge
[16] Egdahl R, Walsh D (1983) *Industry and health care.* Ballinger, Cambridge, Mass.
[17] Goldsmith J (1981) *Can hospitals survive?* Dow Jones Irwin, Homewood, Ill.
[18] United States General Accounting Office (1986) *Medical malpractice: no agreement on the problems or solutions.* U.S. General Accounting Office, Washington, D.C.
[21] Grimaldi PL, Micheletti A (1985) *Prospective payment: the definitive guide to reimbursement.* Pluribus, Chicago
[22] Luft HS (1981) *Health maintenance organizations.* John Wiley, New York
Cowan DH (1984) *Preferred provider organizations.* Aspen, Rockville, Md.
[25] Spiegel A (1983) *Home health care.* National Health Publishing, Owings Mills, Md.
[27] The American Hospital Association publishes a yearly list of multi-institution systems which include multiple hospitals.
[28], [29] [30] American Hospital Association (1987) *Hospital statistics 1986.* AHA, Chicago
[33] Fisk TA (1986) *Advertising health services: what works – what fails.* Pluribus, Chicago
[34] Longo D, Chase G (1984) Structural determinants of hospitals closure. *Med Care* 22: 388
Schatzkin A (1984) The relationship of inpatient racial composition and hospital closure in New York City. *Med Care* 22:379
[39] Hadley J (1982) *More medical care, better health?* Urban Institute Press, Washington, D.C.
Brook RH, Ware JE, Rogers WH et al. (1983) Does free care improve adults' health? Results from a randomized controlled trial. *N Engl J Med* 309:1426–1434

2. The Economic Milieu in Europe: View of the Common Market*

R. HANKIN

Introduction

Within the Member States of the European Community, pharmaceutical consumption is covered by the national health insurance systems. In most cases, therefore, the greater part of the cost is met by the State. Except in the case of over-the-counter products, it is rare for the patient to pay the full cost of a medicinal product. In countries such as the United Kingdom, Germany and the Netherlands, the patient's contribution is limited to a fixed sum irrespective of the cost of the medication, and he is unaware of the actual cost of the product. Even in countries where the contribution is a percentage of the cost of the medication, this share is limited and in general there is little or no patient pressure to reduce prices.

The proportion of national health care expenditures devoted to pharmaceuticals is limited, ranging from 4%–5% in Denmark and the Netherlands to about 10% in France, Germany and the United Kingdom, about 12%–13% in Spain and Italy, and about 20% in Greece and Portugal. However, the combination of an ageing population and high levels of unemployment have resulted in increasing pressures to control costs in all sectors of the social security budget. Expenditures on pharmaceuticals have not been exempt. Indeed, confronted with the hard political choices involved in any attempt to cut welfare expenditures, cuts in pharmaceutical expenditures can appear a relatively easy option, particularly if they are presented as a squeeze on the profits of the industry rather than a reduction in health standards. On the other hand, it is clear that the single-minded pursuit of short-term financial economies will effectively undermine the capacity of the pharmaceutical industry to finance research. At present, therefore, each Member State must balance the objective of controlling public expenditures on pharmaceuticals against the objective of maintaining a competitive research-based pharmaceutical industry. Hitherto, the reconciliation of these two policy objectives has taken place at national level. Some countries, with well-established domestic pharmaceutical industries, have adopted policies which are broadly favourable to the development of pharmaceutical research. Other countries,

*The opinions expressed herein are not necessarily those of the Commission of the European Communities

particularly those more dependent on imports, have tended to give higher priority to the need for savings.

At present there is no consensus on the proper role of the public authorities in regulating pharmaceutical prices or pharmaceutical consumption within the Community. In some countries, such as France, Belgium, Italy, Greece, Spain and Portugal, there is a highly developed system of centralised economic regulation which is strictly applied in the pharmaceutical sector. In other countries, greater reliance has been placed on competitive forces and, in general, public intervention has been limited to an attempt to reinforce competitive mechanisms or to encourage a reduction in overall levels of consumption.

Cost-Containment Measures in Three Member States

In Germany, the national health insurance system is decentralised and it is difficult to impose cost-containment measures from the centre. In any case, such measures would not be in accordance with the post-war German economic tradition. The emphasis has, therefore, been on measures to increase competition and transparency. In an effort to sensitise prescribers to the cost of medication, a special commission has been preparing so-called transparency lists providing an indication of average daily treatment costs for different therapies. Doctors' prescribing habits are monitored and a doctor who prescribes too expensively may face financial penalties. Bavaria has been experimenting with a system of bonuses for doctors who prescribe economically. Steps have also been taken to promote economy at the level of the pharmacist. In several *Länder,* pharmacists are encouraged to dispense a cheaper parallel import from another Member State instead of a more expensive German product, if a parallel imported product is available. An experiment in voluntary generic substitution is in progress in Frankfurt. If the doctor endorses a prescription *"aut simile"* the pharmacist dispenses a generic product and receives an additional fee of 3 DM for the extra work involved.

Finally, in a bid to make further economies, certain categories of medicinal products have been entirely excluded from the scope of health insurance system. In particular, these are products intended for minor and self-limiting ailments.

In the United Kingdom a single ministry, the Department of Health and Social Security, has responsibility for all aspects of pharmaceutical consumption. The department has the dual task of maintaining expenditures on pharmaceuticals at a reasonable level while acting as sponsor to the industry. Reconciliation of these objectives takes place through the Pharmaceutical Price Regulation Scheme (PPRS), which regulates the profits of companies in sales of products covered by the National Health Service (NHS).Each company is assigned a target rate of profit and is free to set the price of individual products, provided it remains within the overall target rate. Price increases must be requested in advance and may be refused. If a company exceeds the profit margins allowed, it may be required to reduce prices and/or repay money to the DHSS, although in certain instances a firm may be allowed to retain unexpected excess profits to reward efficiency and innovation. The target rate of profit is negotiated individually for each firm in the light of its annual financial returns. The system is administratively complex and depends for its success on the establishment of close cooperation between the industry and the department. In 1984–1986 this relationship appeared to come under a certain amount of strain as a result of successive reductions in target rates of profits and disagreements which arose during the renegotiation of the scheme about the amount of information to be provided by multinational companies to justify transfer prices.

Prior to 1985, the PPRS was the main instrument used in controlling NHS pharmaceutical expenditures. In particular, the United Kingdom was unique in that all medicines prescribed by a doctor were covered by the NHS. Over the decade 1975–1985, the relative price of pharmaceuticals in the United Kingdom rose substantially in comparison with other European countries and the government began to look for economies in the budget. Compulsory generic substitution, as recommended in the Greenfield Report, was rejected as unacceptable, as were proposals to follow the German example by entirely excluding certain therapeutic categories from coverage by the NHS. Instead, the government introduced a system of limited lists covering seven therapeutic categories of medicines. These were antacids, laxatives, analgesics for mild or moderate pain, cough and cold remedies, bitters and tonics, vitamins and, perhaps most controversially, the benzodiazepine tranquillizers. The system excludes products from coverage under the NHS if cheaper therapeutic equivalents are available. Moreover, for products which remain available, in most cases doctors are required to prescribe by generic name only; brand name prescribing is prohibited.

Belgium provides an example of a country with very strict controls. These controls operate at two levels. In the first instance, there is a strict system of price control operated by the Ministry of Economic Affairs. No medicinal product may be sold in Belgium until that ministry has approved the price to be charged for the product. The law of 9 July 1975 governing price control establishes general criteria for the overall level of pharmaceutical prices in Belgium and specific criteria for the establishment of individual prices. The general criteria include the overall level of pharmaceutical consumption, the level of health expenditure, the level of prices for consumers, the costs of the different factors of production, promotional expenditures within the industry generally, distribution margins and indirect taxes, research and development investment, the overall profitability of the industry and the health interests of the population. The specific criteria to be considered for each product are also very widely drawn. Taken together they enable the authorities to establish a price for a product on a cost-plus basis, or in comparison with prices in other countries, or on the basis of the therapeutic value of the product in comparison with other products on the Belgian market. Detailed reasons for decisions are rarely given, and once fixed, price increases are difficult to obtain.

In addition, Belgium operates a national formulary, a list of products covered by the national health insurance system.

Once the Ministry of Economic Affairs has approved a price, a company may apply for the product to be included on the positive list of products covered by the health insurance system. Inclusion on the list is subject to strict criteria and can be refused both for lack of therapeutic interest of the product or because its price is too high. Indeed, the health insurance fund, INAMI, can and frequently does demand reductions in the price fixed by the Ministry of Economic Affairs as a precondition to reimbursement.

Taken together, the effect of these measures has been to substantially depress the price of medicines. Between 1975 and 1985 the national resale price index rose from 100 to 191. At the same time, the price of the 50 best-selling products actually declined from 100 to 92.4. Another subject of frequent criticism is the delays inherent in the procedure, which are frequently in excess of 6 months and may run to several years. In these circumstances, there are a number of cases where companies have chosen either not to market a medicine in Belgium or have chosen to market it outside the framework of the reimbursement system.

The Belgian government has itself recognised that such measures may be undermining the profitability of the industry since, in 1985, it introduced a system of *"contrats de programme"* which provided for higher prices for firms entering into a contract with the government to undertake more investment in Belgium.

In fact, the Belgian system of price control has been strongly criticised and is currently under investigation by the Commission to determine whether it is in conformity with the EEC Treaty, both from the point of view of the EEC rules relating to the free movement of goods and from the standpoint of the rules applicable to state aids.

Differences in Price Levels Within the Community

The different types of control which are used by Member States are undoubtedly a factor in the very wide differences in price levels which exist. Although the methodology of price comparisons is controversial, recent studies show clearly that the differential in prices is about 2 ½ times between the cheapest country, Spain, and the most expensive, Germany. Five countries, Germany, Denmark, the Netherlands, Ireland and the United Kingdom have prices above the Community average, while Belgium, France, Greece, Italy, Spain and Portugal have lower than average prices (Table 2.1).

However, other factors are also important. At present the market for pharmaceuticals within the Community is characterised by a remarkable degree of diversity. Like prices, levels of pharmaceutical consumption differ considerably between the Member States (Table 2.1). Germany, with arguably the highest prices in the Community, also has the highest consumption rate, at $ U. S. 90 per capita. However, three countries with low prices also have high consumption rates in money terms: France ($ U. S. 80 per capita), Belgium ($ 67 per capita) and Italy ($ 56 per capita). In contrast, three countries with comparatively high prices have relatively low consumption rates in money terms: Netherlands ($ 35 per capita), Denmark ($ 38 per capita) and the United Kingdom ($ 51 per capita). These figures suggest the existence of considerable variations of consumption in volume terms.

Similar differences are reflected in the statistics available for consumption by therapeutic group. Thus, in money terms, the systemic antibiotics accounted for 20% of Greek pharmaceutical consumption but only 3% of German pharmaceutical consumption in 1982. Medicines affecting the central nervous system, including analgesics and psycholeptics, took 7% of the market in Italy and 22% of the Danish market [4].

It would be difficult, if not impossible, to know how these differences in consumption rates would affect price formation in the absence of controls. However, it does

Table 2.1. Indices of level of prices of medicinal products and per capita consumption

Prices	DE	DK	NL	IRL	UK	BE	FR	GR	IT	ESP	PO
EC Statistical Office, 1983 EC average = 100	169	159	149	118	103	106	78	75	59		
Economists Advisory Group, 1983 UK = 100[a]	140	140	130		100	66	57		66	50	
BEUC, 1987 [1] EC average = 100	152.8		140	142.5	122.6	80.2	68.9		72.3	61	65
Per capita consumption in $ U.S., Economists Advisory Group, 1983[a]	93	51	33	37	48	69	78	36	59	37	25
As a % of GDP at manufacturers' prices, 1983 (Economists Advisory Group)	0.89	0.46	0.36	0.72	0.60	0.86	0.81	1.72	0.96	0.76	1.08

[a] Study [5] published in 1986

seem reasonable to presume that such differences will be reflected in the prices Member States allow to be charged for medicinal products, and, in particular, how an allowance for research and development costs is to be reflected in the price of a specific product.

Parallel Imports

One consequence of the price differentials which exist has been the emergence of the phenomenon of parallel imports, the import of branded pharmaceuticals from low-cost countries and their sale at considerable profit in the high-cost countries. Responsible parallel importing is a legitimate commercial activity but one which gives rise to considerable criticism from the pharmaceutical industry because the consequences of price controls are exported to the high-price countries and funds are diverted away from the research-based industry into the hands of the parallel importer. However, statistics do not differentiate between normal pharmaceutical trade, and the impact of parallel imports is hard to assess. By definition, parallel imports only affect those countries with high prices, and the limited information available suggests that they take about 1% of the German market, under 5% of the UK market and under 10% of the Dutch market. However, parallel importers tend to concentrate on large-selling products, where their market share is somewhat greater. Thus it has been suggested that parallel imports take about 15% of sales of Adalat and Tagamet in Germany (Scrip 27.11.85, p. 5). The number of products affected in this way is probably no more than five or six and in practical terms it is unlikely that parallel importers will be able to obtain sufficient supplies to take a substantially larger market share.

Generics

The impact of generics on the pharmaceutical market constitutes one of the main elements of future uncertainty. Currently no Member State permits generic substitution although, as noted, an experiment in voluntary generic substitution is in progress in Germany, and in the United Kingdom certain products can only be prescribed by generic name. A recent study by the Economists Advisory Group [5] on generic pharmaceuticals in Europe found that generics had a 20% market share in Denmark, 10% in the Netherlands, 6% in the United Kingdom, 4% in Germany and 3% in Italy. In the other Member States it was 2% or less. However, a recent article in *The Economist* gave generics a 15% market share in Germany. This discrepancy is probably due to differences of definition of the borderline between branded generics and "me-too" products.

Prospects for the future expansion of the market share of generics differ from country to country. In principle, they are better in the higher-price countries. However, other factors are also of considerable importance. In the absence of legislation permitting generic substitution, doctors have to be pursuaded to prescribe by generic rather than brand name. In the United Kingdom, the training of doctors has emphasised this for some time. In 1983, the number of prescriptions written generically was about 50% more than the number which could be filled generically, showing a readiness to use generic names even for products still under patent. In Germany, on the other hand, doctors are apparently more concerned about quality and less ready to prescribe generics. Similar reticence prevails in some other Member States.

The attitude of pharmacists is also important. In those countries where pharmacists are paid a proportion of the price of the medicines they prescribe, generic substitution implies a loss of income and efforts to promote generics in France and Belgium have run into problems from this quarter.

The attitude of the national drug regulatory authorities is another factor of crucial importance. The more data a generic manufacturer is asked to provide, the more difficult the marketing of generics becomes. It has now been possible to agree on a common approach to this problem at the European level. For medicines derived from biotechnology and other high technology medicines no abridged application will be possible for 10 years following the authorisation of the original product. There after, the generic manufacturer will be able to submit an abridged application simply comprising full quality data and proof of bioequivalence. For other categories of medicinal products, the waiting period will be at least 6 years and will often be 10 years. These provisions do not affect any patent rights of the original manufacturer.

The Role of the Community Institutions

The existence of such wide differences within the Community clearly poses a significant challenge in the drive for the realisation of a genuine internal market by 1992. It is, therefore, appropriate to conclude with a few words on the role of the Community institutions in this field.

The general principles of the EEC Treaty, in particular the rules on the free movement of goods and the competition (anti-trust) rules, prohibit any discrimination in the operation of national price control systems. The notion of discrimination is interpreted widely: it covers not only formal discrimination but also cases where legislation, which in theory applies equally to domestic production and imports, has the practical effect of putting the marketing of imports at a disadvantage. Where the Commission considers that there is abuse, it may institute proceedings against the Member State concerned before the Court of Justice, and individuals may also challenge the national rules before the domestic courts. Thus the Commission has recently instructed Court proceedings against Italy because the new methodology for controlling prices appears to favour domestic production over imports. The Commission is also extremely concerned about various aspects of the Belgian system and the Belgian government has been invited to submit its final observations before the Commission takes the formal steps necessary to institute proceedings. A communication of the Commission of 4 December 1986 explains in detail how this aspect of the law works [2].

On 23 December 1986 the Commission adopted a proposal for a directive on the transparency of national price control and social security reimbursement measures, its first initiative in this fields [3]. The scope of the proposal was deliberately limited to the establishment of certain procedural rules governing such matters as time limits, the reasoning of decisions and rights of appeal. The underlying objective was to promote a more open and constructive dialogue between industry and the authorities on such matters as the cost-benefit analysis of drug therapy. In addition, the proposal provides for increased cooperation at Community level on two particularly controversial problems, the criteria which should be used to assess the reasonableness of transfer prices and the classification of medicines for reimbursement purposes. The latter is seen as particularly important because our experience in investigating the effects of national systems suggests that, in some cases, medicines are arbitrarily put in one category or another in order to depress the price.

For the future, the Commission will be increasing its activities in this field and will be presenting further proposals to bring national systems into line with the internal market – without seeking to impose a common European price structure or a common European reimbursement system.

References

1. Bureau Européen de l'Union des Consommateurs (1987) Drug prices and drug reimbursement in Europe, a comparative analysis in nine European countries
2. Commission of the European Communities (1986a) Communication from the Commission on the compatibility with Article 30 of the EEC Treaty of measures taken by Member States relating to

price controls and reimbursement of medicinal products. Official Journal of the European Communities No. C 310 of 4.12.1986, p 7
3. Commission of the European Communities (1986b) Proposal for a Council Directive relating to the transparency of measures regulating the pricing of medicinal products for human use and their inclusion within the scope of the national health insurance system. Official Journal of the European Communities No. C 17 of 23.1.1987
4. Economists Advisory Group (1985) The Community's pharmaceutical industry. Office for Official Publications of the European Communities
5. Economists Advisory Group (1986) Generic pharmaceuticals in Europe, blessing or threat? Economists Advisory Group, 35 Albemarle Street, London W1X 3FB

3. The Economic Milieu in Europe: View of the Pharmaceutical Industry

D. TAYLOR

The objective of this chapter is to provide an overview of the key issues related to the production, distribution and consumption of pharmaceutical products in Western Europe. It is not a detailed statistical or comparative review. Rather, it highlights areas of controversy and directs readers' attention towards the broad social and economic factors most likely to influence the evolution of the European-based pharmaceutical industry in the coming decade or so.

The data and arguments presented suggest that although the industry is at present strong and is an important economic asset to most of the major European nations, its environment may deteriorate destructively in the period leading to the close of the century. In part this could be an unintended result of national level measures aimed at controlling overall health spending. In part it might result from international manoeuvring for position within the framework of the European Community (EC).

Avoidance of any such negative developments will hinge primarily on the industry's ability to communicate clearly its true value to the health and wealth of the people of Europe, in both intellectual and emotional terms, and to express reasonably its needs. This must involve industry representatives recognising its faults and weaknesses as well as its strengths and achievements. Yet it will also require independent goodwill and effort on the part of legislators and other influential members of the community, who must be prepared to help find constructive solutions to problems within the overall process of European pharmaceutical supply.

The Pharmaceutical Industry's Contribution to Europe

The Western European pharmaceutical industry consists of over 2000 companies which directly employ about 450000 people. Their annual output is valued at over £ 26000 million (1985), and their exports generate a net trade balance with the rest of the world of some £ 3000 million. Most of this activity is concentrated in the EC countries, which benefit from the presence of 400000 pharmaceutical jobs within their borders and a net positive pharmaceutical balance of trade of about £ 2500 million [1]. The European-based industry spends in total £ 3000 million a year on research and development, principally in West Germany, the United Kingdom, France, Switzerland and Italy [2].

Such data, when combined with an awareness of the direct contribution to community well-being which stems from the pharmaceutical industry's therapeutic innovations, are substantive evidence of the sector's importance to Europe. Nevertheless, the pharmaceutical manufacturers are frequently criticised in relation to issues such as the price/cost of their products, the style and volume of their promotional activities, and medicine safety.

For example, a recent BEUC (Bureau Européen de L'Union des Consommateurs), statement suggested that medicine prices in Germany are 2–2.5 times the Spanish, Portugese, French and Italian levels. [3]. Prices in the United Kingdom, according to BEUC, stand somewhat above the mid-point of the two extremes [3] (1984/86 data). The consumer organisation pointed out that 11 of the 12 EC nations exercise some direct control over pharmaceutical prices, and implied that it would be undesirable if those administrations which most vigorously hold down medicine prices were discouraged from so doing.

Such observations raise interesting questions as to the purpose and possible impact of the currently proposed European directive on price transparency relating to pharmaceuticals. But for the purposes of this opening section, it is more important to stress that simple price comparisons between nations may serve to conceal rather than to reveal the true nature of the pan-European pharmaceutical market. This is in part because of the technical complexities inherent in constructing "baskets" of equivalent goods and adjusting for factors like currency fluctuations [1]. But more importantly, it is because of structural differences in health care demands and processes between nations.

For example, Szuba, in an independent study [4], estimated that despite relatively low price levels, pharmaceutical expenditures – retail level – in both Italy and France in 1983 were actually significantly higher (in the case of France twice) than those of the United Kingdom. His figures suggest three- to fourfold variations in the volume of medicines consumed between such Latin nations and the United Kingdom/United States health care cultures.

Although pharmaceutical industry commentators have indicated some surprise as to the scale of the consumption variations identified by Szuba, the point made is a valuable one. It is interesting to note that despite the widely reported price variations in existence, the great majority of nations in Europe spend 0.6–0.8 of their GNPs on pharmaceuticals, when products' total domestic costs are expressed in manufacturers' price terms net of taxes [5]. That is to say, the lower price, higher volume European countries spend roughly the same amount of their total national wealth on medicines as do the higher price, lower volume cultures. The only notable exceptions to this appear to be the "Scandinavian" model nations like Holland, Sweden and Norway. They combine modest volumes of medicine usage with moderate price levels relative to their national wealth, even though in overall terms their absolute and proportional (percentage GNP) health outlays are high.

The Forces Balancing European Medicine Spending

It would be foolhardy to attempt to construct an elaborate theoretical explanation for a phenomenon which may well prove to be merely a temporary aspect of European health care development. Nevertheless, the consistency of European medicine spending relative to regional wealth demands some general discussion. Three main sets of facts are of importance.

The first point to stress is that (although external observers may tend sometimes to forget) all European countries have long traditions of state intervention in areas like health care provision. Although Britain's National Health Service (NHS) is sometimes held up as a near unique example of "socialised medicine", the historical reality is that the initial British state involvement in health care at the start of the twentieth century was a response to the welfare provisions established in Bismarckian Germany.

The contrast between tax-funded health provisions and those based on compulsory insurance, or between those with higher or lower degrees of overt central direction, or between those which involve higher or lower degrees of consumer choice, should not obscure basic similarities of approach between European nations. Their governments all appear willing and, on occasion, obliged to overide or augment free market forces when it seems to be in the public interest to do so.

The second point is that although health care overall has the characteristics of a luxury good – the richer nations are, the more of their resources they tend to spend on health services – the volume demand for medicines is less straightforward. Naturally "ageing" populations require increased supplies of some forms of pharmaceuticals. But at the same time, people living in more sophisticated health/medical cultures may need and/or consume fewer medicines than their peers in other environments. That is, demand for at least some types of medicines is like that for "inferior" goods.

The third point to stress is that it is reasonable to expect that pharmaceutical research and production in countries with more sophisticated health/medical cultures should be more productive than in less advanced environments. Generally speaking, the larger the mass of the research community in a given nation the more likely it is to generate findings of international significance. Exceptions to this are likely to be countries like Switzerland, which is closely linked with the surrounding European cultures.

Combining these three concept sets, it is apparent that poorer and/or medically less sophisticated European countries will tend to have relatively high medicine demand levels balanced by relatively draconian medicine price controls. Richer and/or more medically sophisticated nations tend to have lower volume usage, but less motive to tightly restrain medicine prices. This is because of both their stronger national pharmaceutical industries and their higher overall health care budgets. Unusual national level tensions and internal conflicts of interest would be expected to manifest themselves in relatively poor but medically sophisticated countries like the United Kingdom. However, if Europe moves more towards being a single market, perhaps with common medicine prices, the policy dilemmas to be resolved will be more complex than those currently faced by any single member state.

The Nature of the Pharmaceutical Market-place and the Basic Options for Cost Control

Right across Europe – and the rest of the world – there are a number of key considerations that policymakers should take into account in addressing the pharmaceutical market-place [6]. They include:

1. The monopsony purchasing potential of national health schemes and of health insurance agencies acting together as purchasing cartels. While there are worldwide governmental policies against undue monopoly (supplier) power, distortions of the demand side of the production/consumption equation seem sometimes to be accepted as desirable.
2. The fact that third-party payment schemes distort normal market relationships, for example, through "moral hazard". This term refers to situations in which consumers who pay a fixed amount for a service feel free, or even perversely obliged, to use it to the maximum possible extent. In addition, "agency" relationships further complicate the demand for medicines. Doctors act as both suppliers and proxy consumers – the need for professional agents is evidence that consumer knowledge is very imperfect.
3. The temporary monopoly power granted by patents. No free market system can support research-based industries without adequate intellectual property laws. Otherwise copyists, who are spared many of the costs of research and development, would undercut innovators' prices. But a long patent term could lead to innovators being able to make an unfair profit or "economic rent", unless their pricing freedom is somehow limited.
4. The cost structure of the pharmaceutical sector with its low (relatively stable) variable costs of production compared to its high non-production costs.

The juxtaposition of variables such as these helps explain why so many European countries have pharmaceutical price/profit control schemes in place. The "need" for these has been particularly accentuated since the oil price crisis of the early 1970s, and the pressure on governments to contain health care costs on the one hand but to improve health care outputs on the other. When this situation is seen in the context of the forces affecting the general balance of European medicine spending (discussed in the previous section) it is not surprising that pharmaceutical cost control emerged as a particularly popular topic for political debate during the last decade.

One well-known contribution to this discussion has been the series of draft papers circulated under the title *Drugs and Money* by Professor M.N.G. Dukes. Each version contains a catalogue of interventions which might be used to limit or reduce pharmaceutical outlays. Logically, they can be divided into two types.

The essential characteristic of the "command" alternatives is one of centrally imposed choice restriction in prescriber and consumer pharmaceutical usage, implemented by mandatory controls. The "market" alternatives leave prescribers/consumers more freedom of choice between different medicines and between medicines and other forms of health care, within the framework of a raised consciousness of the resource implications and "opportunity costs" of their decisions.

From the viewpoint of producers who normally stress their belief in market competition, the "market" option should be more desirable. And from the viewpoint of most

"Command" approaches	*"Market" approaches*
Fixed local budgets for pharmaceuticals	Local budgets for all health care inputs, flexibly allocated through local discretion
Selected (limited) lists of prescribable molecules/ presentations. ("Need clauses" could prevent "unwanted" products from entering the market place)	Greater availability of comparative data, including cost transparency lists
Mandatory generic prescribing/automatic generic substitution	Voluntary generic prescribing encouraged
Use of charges to patients to guide treatment selection – e. g. some medicines free, others charged for	Patient payments related to cost of therapy

of the political leaders of institutions such as the EC, the "command" approach should be an anathema. However, in practice, elements of the "command" approach could in the future prove surprisingly attractive to even the most free market oriented of governments.

A Specific Example – the British Situation

Apart from Switzerland, the United Kingdom benefits more than any other country from its involvement in pharmaceutical trade, relative to its economy's overall size. Britain's 1986 pharmaceutical exports stood at around £ 1450 million, contributing £ 850 million net to the balance of trade. At the same time national volume usage of medicines is relatively low, and the absolute expenditure on all forms of pharmaceuticals (NHS + private/OTC purchases) is about £ 40 per capita (manufacturers' prices). This is well below the per capita expenditure recorded domestically in the other major pharmaceutical trading nations.

The NHS supplies to the public all but a small fraction of the prescritption medicines purchased in the United Kingdom. About 10% of its resources are allocated to pharmaceuticals. This is equivalent to around 0.6% of the GNP.

The profits and costs of manufacturers who supply medicines to the NHS are controlled by the Pharmaceutical Price Regulation Scheme (PPRS) operated by the Department of Health. This has been in operation in roughly its current format since 1969. Its basic structure and operational effects may be summarised as follows:
1. The scheme provides a framework for negotiation between individual companies and the Department of Health and Social Security (DHSS).
2. It confines profits from the sale of branded medicines to the NHS to a current – late 1986 – basic target rate of between 16.5% and 18% on historic capital. Additional

"grey area" profits are permitted to encourage innovation and greater cost efficiency. The proportion of total capital allocated to NHS production is calculated on a sales ratio basis.

3. Other costs, e.g. research, are allowed in relation to the proportion of total company turnover that they account for. Promotion spending is limited by a detailed formula which involves a fixed amount and a turnover percentage, together with product servicing allowances.
4. The current PPRS excludes generic products and introduces mutually agreed arrangements for the confidential monitoring of transfer pricing.
5. The scheme allows a substantial degree of product price freedom, while controlling overall company earnings from the NHS relative to investment and costs. It does not guarantee profits; it ensures that they are confined to certain limits. Companies may well fail to achieve their permissible earnings levels. Thus the scheme ensures that the operation of market forces is a significant determinant of earnings.

Successive British governments have found the PPRS to be satisfactory. It has proved capable of evolving to cope with a changing health care, economic and technological environment, but stable enough to give investors and planners in the pharmaceutical sector reasonable confidence. It has achieved its basic purpose of restraining NHS medicine expenditures without unduly weakening British-based pharmaceutical production and research, or discriminating against other suppliers.

However, the PPRS is not without its critics. Some commentators have urged the introduction of alternative or supplementary cost-control measures. The well-known NHS "limited list", announced in November 1984, and introduced in a revised form in April 1985, was a result of Treasury and related pressures on the DHSS to curb its overall spending. In fact, the limited list is confined mainly to therapeutic areas dominated by products which can be purchased directly by consumers, rather than being obtainable only by prescription. It may have been, to a degree, designed to reduce NHS spending while increasing direct consumer outlays. However, it was certainly intended to encourage health service use of lowest cost alternatives and to cut out of the PPRS calculations altogether certain product costs. The effect of the limited list was to damage the investment confidence of the international industry in the United Kingdom. The £ 75 million per annum saving it is said to have generated for the NHS could very probably have been found in less destructive ways.

Since 1985, the PPRS has been renegotiated and investment confidence in the United Kingdom as a pharmaceutical research and development base has substantially been recovered. Yet there remains, however, a fear in the minds of some industry executives that in the period after the 1987 British general election there may be renewed interest in areas such as NHS list limitation and/or mandatory generic prescribing/substitution. Both, particularly when combined with continuing patent life erosion, could harm considerably the financial position of research-based pharmaceutical manufacturers.

The British situation may be successfully resolved without such undesirable interventions taking place. But from a pan-European viewpoint, it is significant that such debate is taking place within a country to which the pharmaceutical industry is uniquely important. Should the policy debate within the EC focus more specifically around the establishment of a single European pharmaceutical regulation and cost-control system, the British example serves as a clear warning that it could prove difficult for the Community's underlying interest in preserving a strong, research-oriented pharmaceutical sector adequately to be defended.

The short-term interests of less medically sophisticated communities and of governments wishing to check health spending could conceal the longer-term benefits generated by continued investment in pharmaceutical research and manufacture. This paper concludes with a brief discussion of issues relevant to this last area.

The Future – a Single European Pharmaceutical Market?

In analysing the viability of the EC goal of establishing a common European market for pharmaceuticals by 1992, at least three major sets of issues should be examined. All could critically influence the economic milieu in which the pharmaceutical industry may have to conduct about a quarter of its global trade. The first relates to pricing. How important and how desirable is it to try to achieve uniform European medicine prices, as distinct from a greater harmony of principles between the different price control schemes in operation? The second concerns intellectual property, in particular, patent provisions. How far should Europe move to strengthen existing protection for medicines, what benefits might such reform bring, and how can a sensible common policy be determined? And the third centers on regulatory arrangements and the procedures through which pharmaceuticals enter or stay in the market-place.

Regarding this last area, the main debate has centred on whether each nation needs to retain an independent regulatory body, or whether, in time, it might ultimately be preferable for a pan-European agency to be established. The details of the technical discussion are not relevant to this chapter. But it is of importance to note here that some authorities advocate the use of medicine licensing and allied interventions to control the range and cost of products on the market. National systems which already exemplify aspects of this approach include those of France and Norway.

Superficially "need clauses" may sound logical, and could actually prove beneficial to small states which are reliant almost exclusively on others for their new medicines. But for the EC to adopt any such system could prove disastrous for its pharmaceutical sector. It could arbitrarily deny 320 million consumers access to the products of industry research on little more basis than the opinion of a handful of "experts". This would undermine research investment levels and call into question the entire purpose of medicine licensing agencies. Those concerned with protecting the future of the European industry should argue forcefully that medicine licensing is ideally about ensuring the safety and efficacy of products, not about the denial of consumer/prescriber options and the arbitrary limitation of medicine spending.

Turning to the question of intellectual property protection, this is to a degree linked to the regulation topic through the provisions of the "Biotechnology Package". This permits innovators, through the EC authorities, to ensure for a limited period that would-be copyists of certain types of new products will not be able to obtain licences to supply them within the EC. In some ways, such arrangements could be seen as an encouragement the pan-European regulatory approach. However, the protection offered in this context is by no means as comprehensive as that afforded by a full patent. The United States and Japan have moved to provide extended patent terms for medicines which have prolonged pre-marketing development periods. But medicine patent term erosion in Europe is unchecked. The result is likely to be that, unless corrective action is taken, pharmaceuticals entering the American and Japanese markets in the early 1990s will have twice as long effective patent terms as those available in Europe. Should this state of affairs continue for long, the probable result would be the eclipse of European pharmaceutical research.

The opponents of patent law reform and revisions to agreements such as the European Patent Convention appear to base their views on two linked points. They believe that it would be bureaucratically "untidy" and unjustifiable to have special provisions for pharmaceuticals, and that this would, in any case, drive up overall pharmaceutical costs to an unacceptable level. Neither half of this argument is valid.

Pharmaceutical products are a special case, in that their development periods are far longer than any other. And extended patent lives – granted in the context of fair medicine price/profit controls – would not drive up pharmaceutical costs undesirably. Rather, they could protect the financial base of innovators while reducing their dependence on promotional defences. Although it is difficult to model precisely, strengthened patent protection across Europe could permit a better balance of overall cost control and maintained research and development investment while reducing product/research duplication [7]. The sectional pursuit of local interests should not be allowed to deny Europe the collective benefits which a 15–20 year extension of on-the-market medicine patent terms could bring [8].

Finally, in the context of price controls, there are clearly a series of complex problems to be resolved within the EC within the next few years. On the one hand, there are probably benefits to be derived from the creation of a genuinely united common pharmaceutical market for 320 million consumers. A relatively rapid move in this direction would at least lessen the current ambiguities and administrative difficulties associated with the free movement of medicines between EC states, most of which currently operate pharmaceutical price control schemes based on differing criteria. At present, the only groups significantly to benefit from this are the intermediate suppliers, at the expense of primary suppliers and consumers alike.

On the other hand, it has to be recognised that the interests of Member States vary in this area, as do their health care systems and needs. If high-volume consumption communities were to pay the same price for pharmaceuticals (which often have marginal production costs well below their average production cost) as is paid by richer low-volume ones, then their contribution to activities such as research would be unfairly high. Such observations throw the logic of creating any uniform European medicine pricing scheme into doubt. It may be questionable whether the free movement of pharmaceutical goods between European states is genuinely desirable in welfare terms.

References

1. European Federation of Pharmaceutical Industries' Associations (1986) The pharmaceutical industry – a European asset. EFPIA, Brussels
2. Chew R, Teeling Smith G, Wells NEJ (1985) Pharmaceuticals in seven nations. Office of Health Economics, London

3. Scrip No. 1201 1/5/87 p 1. Consumer views on EEC pricing
4. Szuba TJ (1986) International comparison of drug consumption: impact of prices. Soc Sci Med 22:1019–1025
5. Taylor D G, Griffin J P (1985) Does Britain spend too much on pharmaceuticals? Pharmaceutical Journal 23/2/85 pp 228–230
6. Taylor D G (1987) Setting the economic scene. A paper delivered to a DHSS/OHE seminar on the costs and benefits of pharmaceutical research. Unpublished
7. ABPI (1987) Submission of evidence to the House of Lords Science and Technology Committee. ABPI, London
8. Burstall ML, Senior I (1984) The community's pharmaceutical industry. Evolution of concentration, competition and competitivity (Final Report for the European Commission, DG IV). Economists Advisory Group, London

4. Pharmaceuticals and Decision-making in the United States: Cost Consciousness and the Changing Locus of Control

G. R. WILENSKY, L. J. BLUMBERG,
and P. J. NEUMANN

Introduction

Cost-containment strategies have dominated health policy thinking in the United States for much of the last two decades. During that time, annual health care expenditure has grown from $ 42 billion (1965) to $ 425 billion (1985). As a share of gross national product, health care expenditure has increased from 6.5% to more than 10.5%. In the early part of the 1980s, it appeared as though substantial progress was being made in the war against rising health care costs. The annual rate of increase in Medicare[1] hospital expenditure declined dramatically from 20.7% in the early 1980s to 10.2% in 1984, with much of this success being attributed to the introduction of prospective reimbursement for hospitals under Medicare.

Since then, much of the reduction in health care cost increases appears to have been due to the reduction in overall inflation. However, health care expenditure continues to outpace general price increases by a factor of at least 2 to 1. Thus, the success of the changing cost-containment environment of the 1980s is subject to some debate. There is no question, though, that cost-containment strategies in the United States have changed in very profound ways during this period.

The increasing health care expenditure in the United States in the 1970s created enormous pressures for change in the way in which health care is financed and providers are reimbursed. During the decade, many of the cost-containment strategies were centralized and administered at the federal level and relied on federal regulation. One example was the "Certificate of Need" legislation introduced in 1972 which required states to participate in a hospital capital expenditure review process. Another example was professional standard review organizations, introduced at about the same time, which were initially set up as quality control mechanisms for Medicare but primarily served a cost control function. A more severe attempt to contain health care costs in the 1970s involved the Carter cost-containment proposals. The proposals, which were never implemented, were designed to control health care costs through the use of government-controlled fee schedules and other regulatory devices.

---

[1] Medicare is the federal health insurance program for the elderly and disabled

The financial environment of health care in the 1980s can be characterized as a movement toward market-oriented approaches and incentive-based systems. The primary initiatives of the 1980s include the introduction of the prospective payment system by Medicare and the active involvement of employers in health care cost control strategies, including the redesign of employee benefits, the introduction of utilization review and other provider monitoring measures, and the encouragement of alternative delivery systems. The result has been an increased emphasis on market approaches and a competitive system with the simultaneous pursuit of some elements of a major regulatory strategy, namely the price-setting function associated with the Medicare prospective payment system[2]. The cost-containment efforts of the 1980s have forced the various players in the health care arena – providers, payers, and consumers – into more cost-conscious decision-making behavior and have resulted in greater competitive responses by providers at all levels.

These movements have placed hospital administrators and private insurance negotiators in positions of increased authority and power. Many examples of these forces can be seen in decisions involving pharmaceuticals. Physician prescribing behavior has been placed under increasing scrutiny and more stringent controls over prescribing have been instituted. Hospitals and health maintenance organizations (HMOs) are implementing plans to cut pharmaceutical expenses. Pharmacists have more discretion in dispensing drugs and pharmaceutical manufacturers are publishing analyses to show that their products are cost-effective.

This paper examines the change in the nature of pharmaceutical decision-making which has occurred with the onset of competitive influences and cost-containment forces. First, a brief review of the federal agencies involved in assessing and regulating pharmaceuticals is provided, highlighting the increased importance of cost considerations in the regulatory process. A brief discussion of private agencies involved in assessing pharmaceuticals is also provided. The remainder of the paper focuses on how the changing financial environment has affected other levels of decision-making involving pharmaceuticals, including hospitals, HMOs, pharmacists, and physicians.

A Review of Drug/Technology Assessment and Regulation

Before discussing the changing nature of decision-making regarding pharmaceuticals in the United States, it is useful to review the various public and private agencies involved in regulating and monitoring drugs. The more important factors are discussed here along with some historical background, noting especially the increasing attention given to cost considerations in recent years.

---

[2] Under prospective pricing, hospitals are reimbursed according to predetermined rates rather than on the costs incurred in treatment. Payment rates are based on the diagnosis-related group classification scheme which relates patients' diagnoses to the resources used in treatment

The Food and Drug Administration

The Food and Drug Administration (FDA) is the primary regulatory agency for pharmaceuticals in the United States. The agency's responsibilities have increased over the years and now encompass comprehensive safety and efficacy testing.

The first food and drug law, enacted in 1906, stated that labels could not be "false or misleading in any particular" [16]. This requirement was in reference to the composition but not the efficacy of drugs. In 1938 the sale of unsafe drugs was prohibited and specific instructions for appropriate use were required on labels. The 1951 Humphrey-Dunham Amendments classified certain drugs as being legally dispensable only upon a physician's prescription.

In 1962 legislation shifted the focus of drug-related legislation from safety concerns alone to issues of both safety and efficacy [5]. The legislation required that manufacturers demonstrate substantial evidence of the effectiveness of drugs and gave the FDA jurisdiction over clinical testing. The current testing requirements, in place since 1970, stipulate that informal clinical evidence, even evidence verified by experts in the field, is not to be considered an acceptable means for meeting the provision of "substantial evidence of effectiveness" [22]. In 1984 Congress passed "The Drug Price Competition and Patent Term Restoration Act" which extended the effective patent life of drugs and eased the FDA approval process for chemically equivalent generics[3] [6].

The FDA's highly restrictive policies relating to safety and drug efficacy have come under criticism over the years from the pharmaceutical industry, economists, and members of the scientific community. The rigorous clinical trials now required are seen as having two major drawbacks: First, the high cost of the trials, which are presumably passed on to consumers in the form of higher drug prices. Second, the long delays, which may leave patients without potentially beneficial treatments between the time of drug development and FDA approval for market introduction. Of course, less restrictive FDA policies can increase the risk of patients being given harmful pharmaceuticals. Still, many have argued that policies are too restrictive and that some moderation of the approval process is indicated.

The FDA seems to be responding to some of these criticisms. For example, it will soon implement a process allowing patients with "immediately life-threatening" diseases to gain quicker access to drugs still in experimental stages. This shortened process only applies under a treatment protocol/investigational new drug (IND) if the FDA finds that a) the proposed use is intended for a life-threatening condition in patients for whom no satisfactory approved drug or therapy is available, b) the potential benefits of the drug's use are believed to outweigh the risks, and c) there is sufficient evidence of the drug's safety and effectiveness to justify its intended treatment and use [21].

The early introduction of azidothymidine (AZT) is the first example of this limited shift toward speeding the approval process for life-saving drugs. In the urgency surrounding the AIDS crisis, AZT (Burroughs-Wellcome, brand name Retrovir), the

---

[3] Generic drugs are chemically equivalent to brand-name drugs. Unlike manufacturers of brand name drugs, producers of generics usually have little or no costs associated with advertising, marketing, and research and development. Thus, generics can be sold at significantly lower prices.

only drug approved in the United States for treatment of AIDS, was brought to the market prior to the completion of the drug's clinical testing. How much this new policy will affect the introduction of other new drugs remains to be seen. Since the policy is limited to life-threatening diseases with no known effective therapy, its effect will probably be limited.

There is currently no legal basis for the FDA to analyze the cost-effectiveness of pharmaceuticals. However, some in policy circles have suggested that the FDA conduct such analyses. This interest is still largely academic. Adding cost-effectiveness criteria to the already long and cumbersome regulatory process would likely be a problematic and controversial step.

As noted above, a major criticism of the current drug approval process is that it is already slow and costly and inhibits the approval of potentially beneficial drugs. Adding another hurdle to this procedure would increase costs, lengthen approval time, and reduce the number of newly introduced entities. It would also add substantial pressure to an FDA budget that is already severely strained.

In our view, requiring that the FDA determine a drug to be cost-effective prior to its approval is neither a wise nor a desirable position. Because there are a variety of methodological difficulties associated with cost-effectiveness and cost-benefit analyses, the results of such studies are frequently subject to dispute. It is unlikely that one centralized decision-making body can appropriately calculate the cost/benefits or cost-effectiveness for all potential users of the drug. Costs of a treatment are highly variable and specific to individual cases, and benefits are notoriously difficult to measure. Rather than requiring the FDA to determine cost-effectiveness before approving a drug, it may be more advisable to have the FDA simply undertake the analysis and disseminate the results to the public so that consumers and groups can make more informed choices. This information could be enormously helpful to various individuals and groups, such as patients, providers, and third party payers, that may be better suited to decide whether the benefits of particular pharmaceuticals are worth the costs. Another alternative is to have the FDA facilitate the collection of necessary data during clinical trials and make that information available to private groups which would then be free to use it in their own cost-effectiveness analyses.

While cost remains a concern, recent FDA behavior does not suggest that a shift in focus by the agency toward an emphasis on cost consciousness is on the horizon. The AZT case, for instance, is a health care cost increasing decision. The drug prolongs life but does not cure AIDS victims. Use of the drug will probably increase the cost of AIDS, both because of the high cost of the drug itself, and because prolonging life, given the absence of cure, will probably increase the total cost per case.

Medicare and Medicaid Reimbursement Decisions

In addition to the decisions of the FDA, the federal government makes other decisions affecting specific drugs and medical technologies, particularly with regard to reimbursement. Furthermore, private insurers often follow the lead of the Health Care Financing Administration (HCFA), the federal agency responsible for administering the Medicare and Medicaid programs, in making reimbursement decisions. Thus, the federal government's influence extends beyond its immediate sphere.

Medicare and Medicaid (the latter being a joint federal and state health insurance program for the indigent) accept FDA approval as sufficient criteria for in-patient pharmaceutical reimbursement. This is not the case, however, for Medicaid's outpatient drug program. Certain regulatory measures are imposed on outpatient drug costs under Medicaid (Medicare does not currently pay for outpatient pharmaceuticals) by the Maximum Allowable Charge (MAC) program. MAC is a cost-containment not a cost-effectiveness effort. The intention of the MAC regulation is to provide incentives for use of the lowest cost sources available for Medicaid program drugs.

Because HCFA reimbursement for outpatient drugs represents a relatively small part of the outpatient drug market, the administration's attempts to impact pharmaceutical prices has had relatively little effect on the industry. However, if current Congressional proposals to include prescription drugs in the Medicare program are passed, the significant increase in federal purchasing volume that will result will likely provide the government with greater leverage. For example, a bill recently introduced in Congress that proposes Medicare payment of outpatient drugs in excess of $ 500 includes a national formulary that would exclude many high-cost drugs from coverage.

States can also create their own formularies as a way to control pharmaceutical costs in the Medicaid program. Decisions as to which drugs to include in formularies are often implicitly, and at times explicitly, influenced by cost-effectiveness criteria. An example is the well-publicized case of cimetidine, a drug used to treat duodenal ulcers, which was raised in California. Many providers had become convinced that cimetidine was the preferred treatment for patients with duodenal ulcers. However, because of the drug's high cost, Medi-Cal (California's Medicaid program) officials were planning to remove cimetidine from the state's formulary. A cost-effectiveness analysis was used to convince state officials of the relative advantages of cimetidine compared to surgery. As a result, cimetidine remained in the formulary. This action not only affected the availability of the drug to California Medicaid patients, but also set an important precedent for formulary decision-making in the country [10].

The pharmaceutical industry has also begun to realize that cost-effectiveness analysis is an effective strategy that can be used to win HCFA's support on reimbursement issues. Manufacturers now often conduct their own cost-effectiveness analyses and issue reports in support of Medicare, Medicaid, and private insurer reimbursements for particular drug therapies on an outpatient level [2].

Government Centers for Technology Assessment

Comprehensive technology assessment and cost-effectiveness analysis have been justified on several grounds [19]:
1. There is a desire for effective standards for measuring medical quality of both old and new technologies. This is especially helpful for physicians needing assistance in determining effective interventions.
2. Third party payers need guidelines for the use of technology so that reimbursement incentives and policies are structured effectively.

3. Established technologies need to be identified when they become overutilized or obsolete.

Many agencies currently perform some kind of technology assessment. However, the appropriate role of cost-effectiveness analyses in technological assessments has not been well-established. Although the various agencies/organizations described here participate in a range of activities related to drug technology assessment, significant opposition has arisen to the idea of concentrating assessment in a single national agency, and such a move is not likely in the near future.

From 1978 to 1981, the National Center for Health Care Technology acted as a national technology assessment agency. Largely due to opposition from the health care technology industry and some professional organizations, funding for the Center was cut off [14]. In the 1984 Health Promotion and Disease Prevention Amendments, however, Congress responded to a perceived need for centralized technology assessment by mandating that an agency be organized to take over some of the activities of a national center for technology assessment. The Institute of Medicine, an independent, non-profit-making organization chartered through the U.S. government, formed a Council on Health Care Technology as a result of this mandate. The statutory purposes of the council are to promote the development and application of appropriate health care technologies in order to identify obsolete or inappropriately used technologies. The council may appoint an expert panel on coverage and reimbursement of technology issues [15]. The role that cost-effectiveness analysis will play in assessments is still being considered.

The Office of Technology Assessment (OTA) is also involved in issues concerning the cost-effectiveness of many types of technologies, including drugs. Governed by a Congressional Board of six representatives and six senators, it was mandated by Congress to provide "early indications of the probable beneficial and adverse impacts of the applications of technology." Cost-effectiveness, safety, and efficacy assessments are included in this mandate for use in regard to resource allocation decision-making.

The National Institute of Health, the clinical and biomedical research division of the Public Health Service, is also becoming increasingly involved in technology assessment. NIH is interested in medical technologies as they relate to economic, social, and ethical issues. NIH monitors technologies (both new and existing) through use of clinical trials and registries. NIH also holds consensus conferences which produce reports on the efficacy and appropriateness of various technologies.

There are a number of other government agencies which also play a role in technology assessment. The Prospective Payment Assessment Commission, the Congressional Commission which monitors the Medicare prospective payment system, can authorize assessment and cost-effectiveness analyses of pharmaceuticals as such studies relate to issues of Medicare reimbursement. The Veterans Administration Cooperative Studies Program is involved in drug cost-effectiveness analysis and other assessments relating to the specific needs of the veteran population. The Office of Health Technology Assessment (OHTA) at the National Center for Health Services Research also plays a role in the areas of monitoring and evaluating new technologies. OHTA functions not as a national body, but in response to specific requests for technology assessments from HCFA. Cost-effectiveness is a consideration in OHTA assessments.

Private Technology Assessment and Cost-Effectiveness Analysis

Many private organizations are involved, to varying degrees, in technology assessment and cost-effectiveness analysis. In addition to the cost-effectiveness assessment activities at the government level, other private groups participate in this type of analysis [14]. Some examples of these private initiatives are provided here.

Many pharmaceutical manufacturers conduct studies in order to provide evidence of the cost-effectiveness of their products and to promote the use of pharmaceuticals in general. Academic institutions, such as the Institute for Health Policy Studies at the University of California, also conduct cost-benefit/cost-effectiveness studies, as does the Battelle Health and Population Center. In addition, The American Medical Association Diagnostic and Therapeutic Technology Assessment Program examines drug cost-effectiveness insofar as the drugs are applied in medical and surgical procedures.

In addition to other functions, the new Institute of Medicine Council on Health Care Technology is expected to serve as a clearing house for information on technology assessment. Perhaps the Council will be able to fill the existing need to keep track of the types of assessment that are currently being performed both federally and privately. This would help to determine the types of assessment that may be useful but have not been initiated by the many disparate organizations and agencies involved in assessment activities.

Problems with Cost-Effectiveness Analysis for Centralized Decision-Making

Thus far no national consensus on the need for centralized cost-effectiveness analysis has developed. While many agree that cost-effectiveness analysis can be a useful tool which can provide valuable information to policymakers, there are a number of problems which have impeded the establishment of national standards or criteria.

First, it is frequently difficult to establish a consensus regarding new medical innovations. Medical professionals often disagree about the efficacy of therapies, even therapies that have long been used. The dispute over the efficacy and appropriateness of coronary artery surgery versus alternative therapies is often cited as an example [9]. Efforts to reach a consensus on individual therapies are not only very time consuming but also very costly.

Second, methodological problems, such as those described by Wagner in 1983, impose barriers to effective evaluations [23]. Although these barriers may not be insurmountable, they may prove to be expensive to overcome. The problems include:
1. Identifying homogeneous patient groups for analysis
2. Specifying appropriate diagnostic pathways
3. Measuring diagnostic accuracy
4. Measuring diagnostic and therapeutic costs
5. Specifying outcomes of the diagnostic/therapeutic process

Third, because of the inconsistency of technology applications over time and across patients, cost-effectiveness analyses for new medical technologies may be outdated or inapplicable by the time they are completed. As new technologies arise, the indications for the use of the technology being studied change. In addition, as Garrison and Wilensky recently noted, "the problem is further complicated because cost-effectiveness frequently depends on the particular indications of the patient upon whom the technology is applied and the circumstances associated with that application, rather than on the technology per se" [9].

Fourth, the prices and costs needed for cost-effectiveness analysis are difficult to calculate. Acceptable estimates of the costs of treating diseases by alternative therapies are rare. Creating a system to collect such data would be very expensive. Also, unlike other developed countries where drug prices are centrally determined, the prices of drugs and medical devices are not known to the FDA during clinical trials, and prices are not subject to premarketing approval. Even if it were possible to determine prices prior to market entry, these prices would likely change over time. In addition, pharmaceutical indications, dosage regimens, and patient acceptability have significant impacts on cost-effectiveness. All of these factors generally change over time as well.

These problems raise issues of concern when the results of cost-effectiveness analysis are utilized in decision-making. However, these difficulties should not detract from the usefulness of cost-effectiveness analysis when used as one of several pieces of information in assessing new drugs or technologies. In this era of cost consciousness, technological advances, and fears of overutilization, some assessment of health technologies is needed. When financial resources for health care are limited, it is important that health care consumers and purchasers make the most effective purchases possible. Comparisons based on cost-effectiveness analysis provide information about the costs of achieving specific therapeutic results by alternative methods. And within the scope of broader policymaking issues, these evaluations can and should assist decision-makers in addressing questions about how national resources devoted to health might be reduced in a manner that least affects the quality of care.

The Changing Locus of Control and the Focus on Cost Concerns

The increased emphasis on market-based approaches and more serious attempts to contain costs have led to different and more deliberate decisions on several levels. The result has been a movement in the locus of control away from physicians and increasingly toward hospital administrators, private insurers, and pharmacists themselves. At all levels decisions are being made with an attentive eye to the costs involved.

Hospitals

Decision-making at the hospital level regarding pharmaceuticals is changing as a result of changes in the general health care sector. The implementation of reimbursement based on diagnosis-related groups (DRGs) placed a limit on Medicare reimbursement for the treatment of patients. Since inpatient pharmaceutical costs are included within each DRG reimbursement, any reduction in the cost of pharmaceuticals can enhance hospitals' profitability. Similarly, because drugs can replace more expensive therapies, such as surgery, in certain circumstances DRGs may also act as an incentive for the increased use of drugs.

Hospitals have responded to these cost-containment incentives in several ways. Initial action seems to have been most heavily directed toward formulary restrictions. It was hoped that restrictiveness in the creation of hospital formularies would lead to overall pharmaceutical cost savings. However, studies have indicated that although smaller numbers of inventory products result from strong pharmacy product selection policies, there is no significant correlation between formulary restrictiveness and effective cost savings [4]. Costs seem to be influenced by a restrictive formulary only if used in conjunction with good managerial practices.

The use of cost-effectiveness evaluation at the hospital level appears to be increasing. In addition, therapeutic substitution is being used more frequently. Therapeutic substitution is the process in which a different chemical entity, which has similar therapeutic properties but is less costly, is chosen for the treatment of patients. Hospital pharmacy and therapeutics (P & T) committees have taken the lead in initiating such investigations. Evaluations of frequently used drugs such as aminoglycosides, cephalosporins, and methylprednisolone have been performed repeatedly. The results of these studies have consistently shown cost savings from therapeutic substitution [1, 3, 11, 12, 20].

Cost-effectiveness analyses and therapeutic substitution strategies also seem to be gaining acceptance, after some initial resistance, by physicians within the hospitals. P & T committees often turn to in-house or externally performed cost-effectiveness analyses for assistance in making appropriate substitution policies.

In the case of aminoglycosides, some P & T committees hypothesized that the more expensive tobramycin could be replaced with gentamicin in an effort to reduce costs. Multiple studies were performed at the individual hospital level that demonstrated cost savings and substitutability between the two antibiotics [3, 11, 12]. Similar studies were done in order to evaluate the cost-effectiveness of first-generation cephalosporins in place of the more expensive second-generation cephalosporins [1, 12, 21]. Again, the studies showed the potential for significant cost savings with comparable outcomes.

Computers have assumed a major role in cost analysis and have enhanced the ability of hospitals and pharmacies to use cost-effectiveness analyses [18]. Programs have been created for use by P & T committees that input data on specific antibiotics as well as hospital-specific data on the costs of personnel and supplies. The output is a cost comparison between alternative antibiotic regimens. However, such programs can be limited in their ability to account for varying efficacy and toxicity, and for indirect costs and benefits such as potential side-effects and patient satisfaction.

Strategies for influencing in-house prescription dispensing habits in cost-saving directions have taken various forms. One approach has been a voluntary educational effort. Other more restrictive approaches have required physicians to obtain a specialist's approval before prescribing nonformulary pharmaceuticals.

A study conducted at the University of Minnesota Hospital, a 735-bed tertiary-care hospital, involved a voluntary approach [8]. A clinical pharmacist and infectious disease physicians developed information sheets for physicians which described the appropriate use of the target drugs vancomycin and tobramycin. The clinical pharmacist would contact a physician when he/she prescribed drugs for indications that did not meet the established criteria. During the 1-year study, changes in the drugs' utilization resulted in a net decrease of $ 161 396 in drug expenditure. The program demanded the time of 0.5 full-time employees, and the return for the service exceeded a ratio of 10:1.

Another study conducted at the Downstate Medical Center in Brooklyn, New York, found that 31% of total antibiotic expenditure at the hospital was attributable to cephalexin (a cephalosporin) [20]. In a first effort to decrease the use of cephalexin in favor of a lower cost cephalosporin, physicians were required to complete an antibiotic justification form when prescribing the drug. While this requirement alone did not decrease use of the drug, marked drops in cephalexin utilization occurred when physicians were required to telephone an infectious disease physician for approval prior to prescribing. The study team concluded that relatively little effort was demanded on the part of the infectious disease expert, and total antibiotic costs fell 29%.

Other hospital pharmacies have used notices to inform physicians that specific therapeutic substitution policies have been adopted, and that physician prescriptions for particular drugs will be routinely filled with alternative drugs. A drug utilization review committee in one particular hospital projected first year savings after policy implementation at over $ 139000 in patient charges [11]. A 72% rate of compliance with the policy was demonstrated.

Strategies directed at hospital pharmacists have also been developed to encourage closer attention to, and compliance with, P & T committee conclusions drawn from cost-effectiveness analyses. An example is a hospital-wide monetary incentive program at United Hospital Incorporated and Children's Hospital Incorporated in St. Paul, Minnesota [13]. The program was developed under the assumption that directly rewarding pharmacists for efforts to control costs (through therapeutic substitution and other efficient managerial practices) would reinforce the effects of other cost control programs, including formulary control and improved purchasing and inventory control. In 1984, the decline in drug and IV expenses per adjusted case led to an average payment of $ 500 to each full-time pharmacist.

To the extent that the hospital pharmaceutical strategies described here entail pharmacists making routine replacements of physician-prescribed drugs, decision-making control is transferred from physicians to pharmacists and P & T committees. In important ways, the decisions that hospital physicians are still making are changing as well. Direct caregivers are now being forced to justify therapeutic choices as being cost-effective. The cost of treatment was a much less significant consideration in the past.

While reducing costs have been the focus of many new strategies, the impact of these strategies upon quality of care has received less attention. An important question that will need to be addressed is whether cost-containment programs are affecting the quality of patient care. Some monitoring or review will be necessary in order to insure that patient care is not declining because of unwarranted therapeutic substitutions or other cost-containment measures.

Health Maintenance Organizations

Because health maintenance organizations (HMOs) provide medical services to enrollees in return for a fixed prepaid capitation payment, they have a financial incentive to minimize the cost of all services. HMOs usually include prescription drug costs (both inpatient and outpatient) in their capitated fees. HMOs frequently maintain their own pharmacies or provide for capitated pharmacy services with little or no associated cost-sharing. In many cases, stringent cost-containment policies have been implemented at the pharmacy level. Because of this, the pharmaceutical industry and consumer advocacy groups have expressed fears that drugs have been underutilized and that quality has suffered as a result of such internal regulation. However, to date there is little evidence that would substantiate these fears.

A recent study by Lewin and Associates found that HMO cost-containment measures generally lead to three basic organizational policies [7]:
1. An increase in formulary use
2. Growth in use of least-cost generic drugs
3. Pressure on physicians to prescribe as little and as inexpensively as possible

In addition, the Lewin report indicated that HMOs attempt to discourage their physicians from meeting with representatives of pharmaceutical manufacturing companies and that 85% of HMOs encourage the dispensing of therapeutic alternatives. The report also stated that HMO physicians were less apt than their fee-for-service counterparts to support research-based pharmaceutical companies and their products.

The Lewin study suggests that physician decision-making regarding pharmaceuticals is circumscribed in HMOs and that cost considerations play a major role in the decision process. The report, however, did not attempt to examine the actual quality of patient care. It remains an open question whether or not the decision-making processes in HMOs have actually had an adverse impact on the quality of patient care. Presumably, HMOs' interest in maintaining a long-term relationship with enrollees will influence their decisions in a way that will keep them quality conscious as well as cost conscious. Reducing needed or appropriate pharmaceuticals could result in greater long-term medical expenses and in unsatisfied customers who may enroll in other plans.

Although HMOs have incentives to maintain high quality and provide appropriate levels of care, due to competition for patients. There has still been concern about the potential for underutilization of necessary services. Critics note that because enrollees have imperfect information and frequently cannot understand the complexities involved in medical treatment, there is a need to monitor the care provided in HMOs. This concern is particularly acute for the elderly. Various systems are being planned which will monitor quality of care in HMOs, particularly for outpatient services. Systems may, for example, monitor the appropriateness of HMO formularies, the appropriateness of physician prescribing patterns, and the availability of new technological innovations for specified diagnoses.

Pharmacists

The role of the individual community pharmacist has been changing as well in recent years. The movements toward the use of generic equivalents and therapeutic substitution have given pharmacists more authority and decision-making responsibilities. Increasingly, pharmacists are being pressured to select the most cost-effective drugs. Pharmacists are also finding that to maintain market competitiveness, they must participate in "community" buying groups. The advantage to individual pharmacies of bulk buying comes from the buying groups' access to the lowest cost drugs available and easy access to a diverse and extensive supply of pharmaceuticals.

Therapeutic substitution and the growing emphasis on cost-effective drug therapies are also issues of concern for the community-based pharmacists. Therapeutic substitution was the subject of a 1985 conference sponsored by Medicine in the Public Interest, Inc. [17]. The participants of the conference, which included physicians, pharmacists, representatives from the pharmaceutical industry, and academics, expressed concern about the use of therapeutic substitution in the community pharmacy setting. In general, both physician and industry concerns on issues of therapeutic substitution and the use of cost-effective drug therapies were much greater for the uncontrolled community environment than for the hospital setting. There were also some other specific concerns. First, there was some concern that some pharmaceutical cost increases will occur as a result of pharmacists' expectations of reimbursement for exercising independent authority. Second, fears existed that as the market for pharmaceuticals shrinks because of widespread substitution, and as less innovative drugs become less marketable, large price increases will occur for new "blockbuster" drugs so that manufacturers will be able to recoup their research and development costs with a smaller number of product introductions. Third, some participants felt that pharmacists' lack of knowledge of patient histories and requirements could lead to suboptimal or even harmful substitutions unless communication between physicians and pharmacists was strengthened.

Physicians

The increase in administrators' and pharmacists' decision-making authority has been accompanied by a loss of power on the part of physicians. Physicians today have less discretion in prescribing drugs, and where they retain discretion, they are being forced to be more cost-conscious in their prescribing patterns.

An issue which has recently received significant attention and continues to be hotly debated is physician dispensing of pharmaceuticals. More widespread dispensing could return some pharmaceutical decision-making control to physicians. This type of service would allow physicians to determine whether brand-name drugs would be prescribed more often than generics. It would also give them the control over any potentially indicated therapeutic substitutions. Physicians who dispense pharmaceuticals from their offices would, however, be forced to keep a cost-conscious eye on competition in the pharmaceutical dispensing marketplace. Based on both economic and quality grounds, there has been pressure from pharmacy and consumer groups to ban physician dispensing except under limited conditions. Among the concerns have been a number of important ethical and practical considerations:

1. The incentives for physicians to overprescribe
2. The potential for physicians to prescribe drugs that might not be appropriate but that he/she has available
3. The potential for physician support staff to do the actual dispensing, increasing the likelihood of errors

Physician dispensing will be an emotional and controversial issue in the near future. However, the full impact of dispensing is uncertain partly because it is unclear how the market for health services will evolve. If, as some suggest, there is a continued movement toward "one-stop shopping" with clinics and managed care plans offering physician, outpatient surgery services, and pharmaceuticals under one roof, the lines of decision-marking authority may be blurred further.

The Pharmaceutical Manufacturing Industry

How significantly the changing locus of control and the increased focus on cost issues in pharmaceutical decision-making will affect innovation in the drug industry remains to be seen. One would expect, however, that companies involved in extensive research and development are realizing that their best market opportunities lie in the development of cost-reducing and cost-effective products. Companies are already performing cost-effectiveness analyses of their own drugs as a marketing strategy. Even though research and development dollars are growing steadily (pharmaceutical price increases appear to be facilitating this), new drug approvals by the FDA declined from 30 in 1985 to 20 in 1986. It is possible that as the industry focuses its attention on the production of cost-effective pharmaceuticals and as the costs associated with the approval process climb, the number of new drugs will continue to decline. Corporations may consider it unprofitable to invest in the development of a new drug and to submit it to the expensive approval process if they feel that they will be unable to show that the drug is cost-effective. To the extent that this results in more research aimed directly at the development of drug therapies which can substitute for more expensive and/or invasive therapies, society will benefit. However, to the extent that it dissuades the industry from investing in potentially valuable research and development because of fears that the product will not be shown to be cost-reducing or cost-effective, society will suffer.

Conclusion

As decision-making in health care has become more a function of market-based incentives, the control over and the criteria involved in pharmaceutical decisions have shifted. Whereas regulation at the federal level by the FDA and control by individual physicians at the prescribing level have historically been the focus of decision-making with only limited concern for costs incurred, the advent of a more competitive system has transformed hospitals and pharmacies into the sources of new decisions based on considerations of cost.

Health care delivery in the United States is changing at a rapid pace. As financing pressures persist, provider roles and decision-making structures will continue to change. The trade-offs between reducing costs and improving quality of care will increasingly be of concern. These trade-offs will be felt at every level of pharmaceutical decision-making as attention continues to focus on controlling costs.

References

1. Abramowitz PW, Nold EG, Hatfield SM (1982) Use of clinical pharmacists to reduce cefamandole, cefoxitin, and ticarcillin costs. Hosp Pharm 39:1176–1180
2. Balinsky W (1986) Home care prescription drug reimbursement: a case for intravenous antibiotics. Hoffman-La Roche Public Issue Report. Nutley, New Jersey
3. Cynamon MH, Kallet KK, Zacher DE (1982) Aminoglycoside policy directed at use and cost. Hosp Formul 17 (Jun):811–814
4. Daniels CE, Wertheimer AI (1982) Analysis of hospital formulary effects on cost control. Top Hosp Pharm Manage 2 (Aug):32–49
5. Drug Amendments Act of 1962, 87th Congress, 2nd Session, 10 October 1962
6. The Drug Price Competition and Patent Term Restoration Act, 98th Congress, 2nd Session, 23 January 1984
7. FDC Reports ("The Pink Sheet"), vol 49, No. 8, 23 February 1987
8. Fletcher CV, Giese RM, Rodman JH (1986) Pharmacist interventions to improve prescribing of vancomycin and tobramycin. Am J Hosp Pharm 43:2198–2201
9. Garrison LP, Wilensky GR (1986) Cost containment and incentives for technology. Health Affairs 5(summer):46–58
10. Geweke J, Weisbrod BA (1982) Clincical Evaluation vs. Economic Evaluation: The Case of a New Drug. Medical Care, Vol. 20, No. 8: 821–830
11. Guernsey BG, Berina LF, Lazarus MC et al. (1985) Cost containment through therapeutic substitution of aminoglycosides. Hosp Pharm 20:82–83, 86–88, 90–92, 95
12. Hayman J, Sbravati EC (1985) Controlling cephalosporin and aminoglycoside costs through pharmacy and therapeutics committee restrictions. Am J Hosp Pharm 42:1343–1347
13. Herrick JD, Manning SH (1985) Monetary incentive for pharmacists to control drug costs. Am J Hosp Pharm 42:1527–1532
14. Institute of Medicine (1985) Assessing medical technologies. National Academy Press, Washington, D. C.
15. Institute of Medicine (1986) Newsletter of the Council on Health Care Technology, National Academy of Sciences, vol 1, No. 1. Washington, D. C.
16. Lashof JC (1982) Medical technology decision making in a political environment at the Federal level. In: Bloom BS (ed) Cost benefit and cost-effectiveness analysis in policy-making: cimetidine as a model. Biomedical Information Corporation Publications, New York
17. Medicine in the Public Interest, Therapeutic Substitution Symposium, December 1985, Boston, Massachusetts
18. Parr MD, Hansen LA, Waite WW, Rapp RP (1986) Computer program for comparing total costs of intravenous antibiotic regimens. Am J Hosp Pharm 43:2189–2193
19. Roe W, Anderson M, Gong J, Strauss M (1986) A forward plan for Medicare coverage and technology assessment, vol. II. Supporting documentation. Prepared for the Assistant Secretary for Planning and Evaluation, Department of Health and Human Services. Lewin and Associates, Washington, D. C.
20. Seligman SJ (1981) Reduction in antibiotic costs by restricting use of an oral cephalosporin. Am J Med 71:941–944
21. Sorian R (ed) (1987) Medicine and health. McGraw-Hill, Washington, D. C.
22. Temin P (1980) Taking your medicine. Harvard University Press, Cambridge, Mass.
23. Wagner JL (1983) The feasibility of economic evaluation of diagnostic procedures. Soc Sci Med 17:861–869

5. What Price Pharmaceutical Innovation?

P. ZWEIFEL

Introduction

The objective of this contribution is to explain how pharmaceutical innovation comes about and what its effects are in terms of both private and social costs and benefits. The paper will be divided into three parts. First, the incentives of a pharmaceutical firm to innovate are analyzed. Starting from the new theory of demand as developed by Becker [1] and Lancaster [6, 7], quality competition through product innovation is compared to price competition through process innovation. This theory can also be used to shed light on the pricing decisions of innovators and on how price regulation influences the balance between quality and price competition.

The second part of the paper is devoted to an analysis of private costs and returns associated with product innovation. Pharmaceutical research and development progresses through six phases, each of which leads to the abandonment of many projects. Because the flops among the projects cannot be identified initially, the costs of a successfully launched product should be estimated from data relating to an entire *cohort* of would-be innovations. On the basis of such cohort data from the United States, estimates of the real internal rate of return to pharmaceutical innovation are derived and compared with that of product innovation in other industries.

In a final step, the transition from a firm's private costs and benefits to their social counterparts is made. This transition is particularly difficult in the case of the pharmaceutical industry because it sells its products largely in an *insured* market. This means that conventional estimates of social benefits, which are based on observable market quantities, overstate true individual willingness to pay. An approximate method of correction for this insurance-induced distortion is presented, resulting in lowered social rates of return of pharmaceutical innovation, which, however, still tend to be comparable to those of other industries. The existence of such distortions may lead one to advocate valuation of pharmaceutical innovation by other agents, in particular the insurer or a Government agency. However, it will be shown that their willingness to pay quite likely suffers from substantial biases too.

The contribution will be rounded off by some conclusions regarding the nature and social value of product innovation in the pharmaceutical industry.

Pharmaceutical Innovation as Product Innovation

Outside of the health care sector, technological change is the key to ever rising productivity and lower costs. For example, the progress achieved in electronic data processing has been comparable to a change resulting in a Cadillac of today costing U. S. $ 500 instead of U. S. $ 50000 30 years ago (extrapolating from Chow) [2]. In stark contrast to this very favorable view of technological change, the financiers of health care, insurers, and Government agencies alike fear medical and pharmaceutical innovation because of their cost-increasing impact. This puzzle calls for an explanation.

From an economic point of view, there may well be a disturbed balance in the health care sector between process innovation and product innovation. Whereas process innovation results in an unchanged good or service being produced at lower cost, product innovation changes existing characteristics of the product or adds new quality dimensions to it. For the management of a company, process innovation means changing job contents of workers in a continuing effort to hold production costs down. This is a rather unpleasant task, creating considerable potential for conflict. Product innovation on the other hand opens up new opportunities in research and development (R+D), marketing, and production. Thus, one would expect a natural preference for product innovation over process innovation among the work force of a firm. It is the buyer of the final product or service who, by balancing higher quality against higher price, forces producers to exert their efforts in favor of process innovation. Indeed, the low quality product may often be the better buy from the point of view of the consumer.

Figure 5.1 demonstrates this assertion. Let there be a drug featuring two characteristics, freedom of pain (c_1) and therapeutic effect (c_2). For example, drugs presently available for the treatment of rheumatic diseases are quite effective as pain killers (ranking high on the c_1 dimension) but do not change the mechanisms causing the erosion of joints (ranking very low on the c_2 dimension). For simplicity let there be a traditional drug, a unit of which (one daily dose, say) leads to an amount of pain reduction as marked by point 1T on the c_1 axis of Fig. 5.1. Point 1T is a technological description of the drug and has no relationship to cost of production or price whatsoever [6]. Accordingly, a fully insured patient (and his physician as an agent) will always prefer a new, more effective drug. This new pain killer is marked as point 1N on the c_1 axis of Fig. 5.1. This preponderance of technological aspects has a very important consequence for the pharmaceutical firm: Given *full* insurance, it always pays to launch a new, more effective drug. This line of reasoning may be summed up in

---

Conclusion 1
If the patient is fully insured, he will always prefer the more effective drug, thus encouraging product rather than cost-reducing process innovation by pharmaceutical firms.

Things may change radically if the patient is not insured or if he has to pay at least part of the price of the drug out of pocket. For simplicity, assume that he is not insured at all and that a certain budgeted amount of money would suffice for four doses of the traditional drug. This patient is therefore able to enjoy the services of drug T during

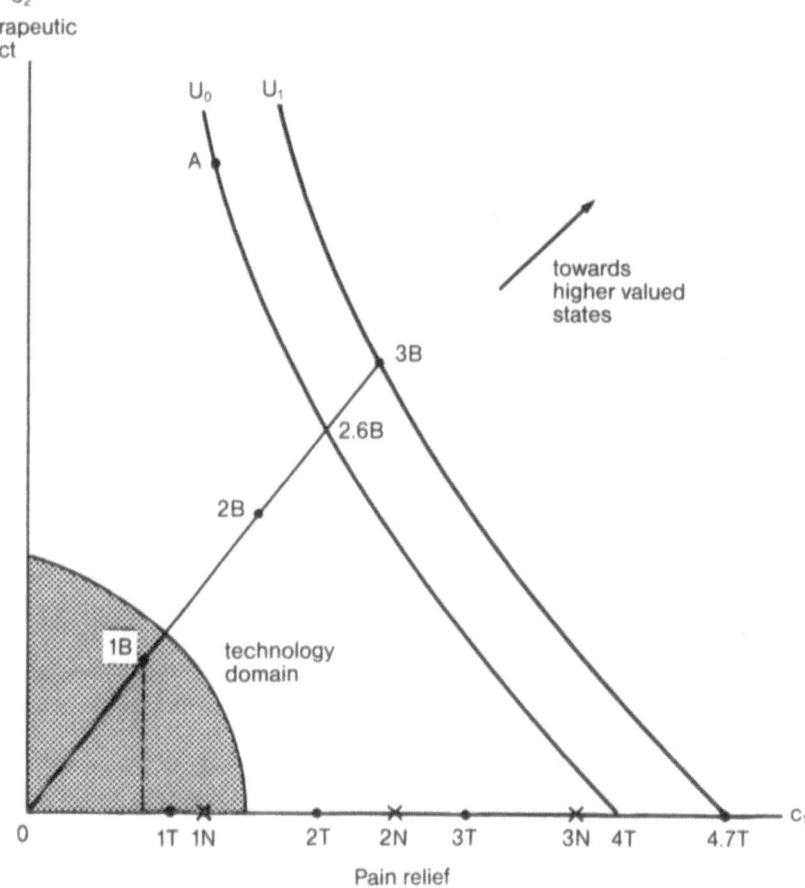

Fig. 5.1. Price and quality competition between different products

four periods, which transforms the technological point 1T of Fig. 5.1 into the economically relevant point 4T. Additionally, assume that the new drug is more expensive
so that the same budgeted amount will buy only three doses of it. Thus, the relevant
point for the consumer's decision becomes 3N on the c_1 axis [see 7 Chap. 3]. In this
particular example, total services rendered by drug T exceed the amount rendered by
drug N, although N is technologically superior. In fact, the old drug is able to win over
its new competitor because its lower quality is more than offset by its lower price.
Generalizing somewhat, we may state

Conclusion 2

The greater the patient's share in the cost of the drug, the better are the chances
of the less effective but lower cost drug being preferred, and the greater the
incentive for pharmaceutical firms to pursue cost-reducing process innovation.

A *breakthrough* innovation can be defined as one resulting in a product or service that contains quality dimensions not available before. Inventions such as flexible yet durable fibres (nylon), focused instead of dispersing light beams (laser), or the typewriter with lift-off correction facility are examples in point. In our context, it may not be too unrealistic to assume that in the foreseeable future there will be causal antirheumatic drugs having a therapeutic effect, resulting in a permanent improvement in the status of rheumatic joints.

In terms of Fig. 5.1, such a drug would be located away from the c_1 axis because it also produces a positive therapeutic effect ($c_2 > 0$). Let its technological quality be represented by point 1B in Fig. 5.1, indicating that this chance of permanent improvement goes along with a reduced effectiveness in pain relief in comparison with both preparations T and N (cf. points 1T, 1N, and the projection of point 1B on the c_1 axis).

In such a case, the regulatory agency in charge of marketing approval faces a dilemma. How is it to trade off this new characteristic "therapeutic effect" against the loss in terms of pain relief? It is quite likely that the innovator would not be allowed to introduce his product into the market because it fails to satisfy quality standards set by brands already on the market. Such a decision may well hurt an important group of patients, however. Their loss can be shown by introducing the subjective valuation of the patient concerned as an additional element into Fig. 5.1. The subjectively most preferred state would be represented by points lying as far as possible away from the origin, indicating full pain relief from pain as well as normal functioning of joints. Such a "bliss point" can be envisaged like the peak of a mountain, the height of which mirrors the degree of subjective desirability of a state. But this implies that there must be curves of equal altitude around that peak.

One particular curve of equal "utility" is shown as U_0. It suggests that the patient considered would put the same value on almost complete pain relief at point 4T (as produced by four doses of drug T) and another state like the one depicted by point A, where he would suffer a fair amount of pain but have a good chance of permanent improvement. Before the advent of the breakthrough innovation, the patient depicted in Fig. 5.1 could do no better than spend his drug budget on four doses of drug T and reach a level of utility as given by U_0. Due to the breakthrough innovation, point 3B becomes available for the patient, assuming that this new drug is as costly as the previously launched minor innovation (i. e., the budget of the patient buys three doses, as before). Point 3B lies on curve U_1, which is clearly higher valued than U_0. The distance between the two curves U_1 and U_0 mirrors the size of the loss inflicted on that patient by a ruling that drug B is not to be marketed. Interestingly, this loss can even be expressed in terms of *money*. With drug B available, the patient reaches point 3B on curve U_1; this is subjectively equivalent to some 4.7 doses of T. Without drug B, he is confined to point 4T on curve U_0. Thus his loss amounts to 0.7 doses of the traditional drug T, which can be expressed in money given the price of T. Therefore, if the shape of the curves U_0 and U_1 is (approximately) known, the equivalent monetary value of the lacking availability of drug B can be calculated. In all, we state

---

Conclusion 3

Modern demand theory suggests that a new drug featuring additional desirable qualities over existing alternatives has a competitive advantage that can be expressed in terms of money. Barring such an innovation from the market because of weak performance in other dimensions of quality will thus inflict losses on at least subgroups of patients.

The distance between curves U_1 and U_0 in Fig. 5.1 also indicates the leeway the innovator has for his pricing policy. Indeed, the producer of drug B could raise its price while still remaining competitive with regard to drug T. The maximum possible price increase can be read off in Fig. 5.1 as the distance between points 2.6B and 3B, i.e., the equivalent of 0.4 doses of B. In relative terms, it amounts to 0.4B/3B or 13%. After this price increase, the patient could either spend his budgeted amount on 4 doses of T or 2.6 doses of B to be equally well off. More generally, the innovator who is able to satisfy a quality dimension over and above competing brands can charge a higher price even though his product may be inferior in terms of quality dimensions shared with competitors [see 7, Chap. 5].

It should be noted that this leeway in pricing obtains only if the innovator is permitted to introduce his product in the market. This means that above all, product characteristics must satisfy the demands of the regulatory authority. To the extent that a regulatory decision on technological norms is predicated on existing rather than newly offered quality dimensions, *product design* becomes of overriding importance. Thus, the existence of authorities regulating market launch (such as the Food and Drug Administration in the United States) serves once more to stifle cost-reducing process innovation while encouraging product innovation. This argument leads to

---

Conclusion 4
Besides the effects of health insurance, the presence of governmental agencies regulating market launch serves to decrease the importance of price compared to quality and hence of process innovation in favor of product innovation in the pharmaceutical industry.

The Development of a New Drug

In the preceding section, product innovation just happened. In this section, emphasis is shifted to the process that leads to the creation of a new drug. This process can be seen as one particular form of investment by the firm competing with other possible investments, from building new plants to buying securities.

The allocation of scarce funds among competing investment projects immediately raises the question of the respective rates of return that can be reaped.

Basically all investment projects compete with just putting the money in a bank. But they may well differ between each other as to their time structures of outlays and returns generated. Discounting to present value renders these different projects comparable. If put in a bank at a rate of interest of 5%, 1 dollar will be 1.05 dollars 12 months later. Actual rates of interest are often much higher because they also contain a premium for expected inflation. This premium will be neglected for simplicity in the following; instead, the focus will be on the so-called real rate of interest. But if 1 dollar today in the bank is 1.05 dollars 1 year later, this also means that 1.05 dollars receivable with certainty in 1 year has the same present value as 1 dollar receivable now. Put another way around, 1 dollar receivable 1 year hence has a present value of 1/1.05 = 0.952 dollars today. Therefore, at a real rate of interest of 5%, both costs and returns associated with an investment project that accrue with 1 year's delay are equivalent to 0.952 dollars at present value. If the delay is 2 years, present value falls to $0.907 = 1/(1.05)^2$ dollars, and if 10 years, to $0.613 = 1/(1.05)^{10}$. Thus, deferred costs and returns have to be discounted. These considerations give rise to

---

Conclusion 5

The discounting procedure serves to make flows of outlays and revenues differing in their distribution over time comparable. The greater the delay and the higher the real rate of interest, the greater the impact of discounting, reducing present values of both outlays and revenues.

Product innovation in the pharmaceutical industry is characterized by two salient features. First, returns typically flow with a delay of some 10 years. For the development of a drug prior to marketing goes through six phases of research and development: A discovery phase is followed by preclinical tests performed on animals, with the objectives of identifying toxic effects and establishing the main pharmacological characteristics of the new substance. The next three phases are devoted to clinical testing, while the sixth and last phase consists of testing for long-term toxic effects.

The second feature of the pharmaceutical research and development (R+D) process is the high degree of risk. While the R+D process is under way, costs accumulate, and most of the outlay is caused by projects that have to be dropped. Hansen [4] studied the rate of attrition among a cohort of some 1000 projects during these 10 years prior to market launch. He found that only one out of eight would-be innovations survives the process, i. e., 88% of them are cancelled sooner or later. Figure 5.2 shows the typical outlay profile of a portfolio consisting initially of eight substances discovered in 1967. In 1977, one of these eight products would have been introduced onto the market.

Figure 5.2 also shows the return phase of the marketed new drug, based on cohort estimates by Virts and Weston [12]. Initiated in 1967, patent protection for this successful innovation would have expired by 1986.

Prior to discounting, the average pharmaceutical innovation looks like a good investment because its (pretax) returns increase very quickly after market launch, reaching a value of 6 million dollars (annually) at 1967 prices. Since this curve shows costs incurred and returns reaped during a given year, total costs and total returns can be read off from the area under the curve. The comparison of the "return area" and the "cost area" suggests that a handsome profit can be made from pharmaceutical innovation. But when the same comparison is made using a 5% rate of discount (say) for calculating present values, the "return area" shrinks considerably because cash flows are deferred by at least 10 years. As a matter of fact, pharmaceutical innovation (as a member of the 1967 cohort) is only slightly more profitable than a financial asset yielding a real rate of interest of 5%.

Alternatively, one can also ask the question, "What would be the interest rate earned on money in the bank that would turn pharmaceutical innovation from a

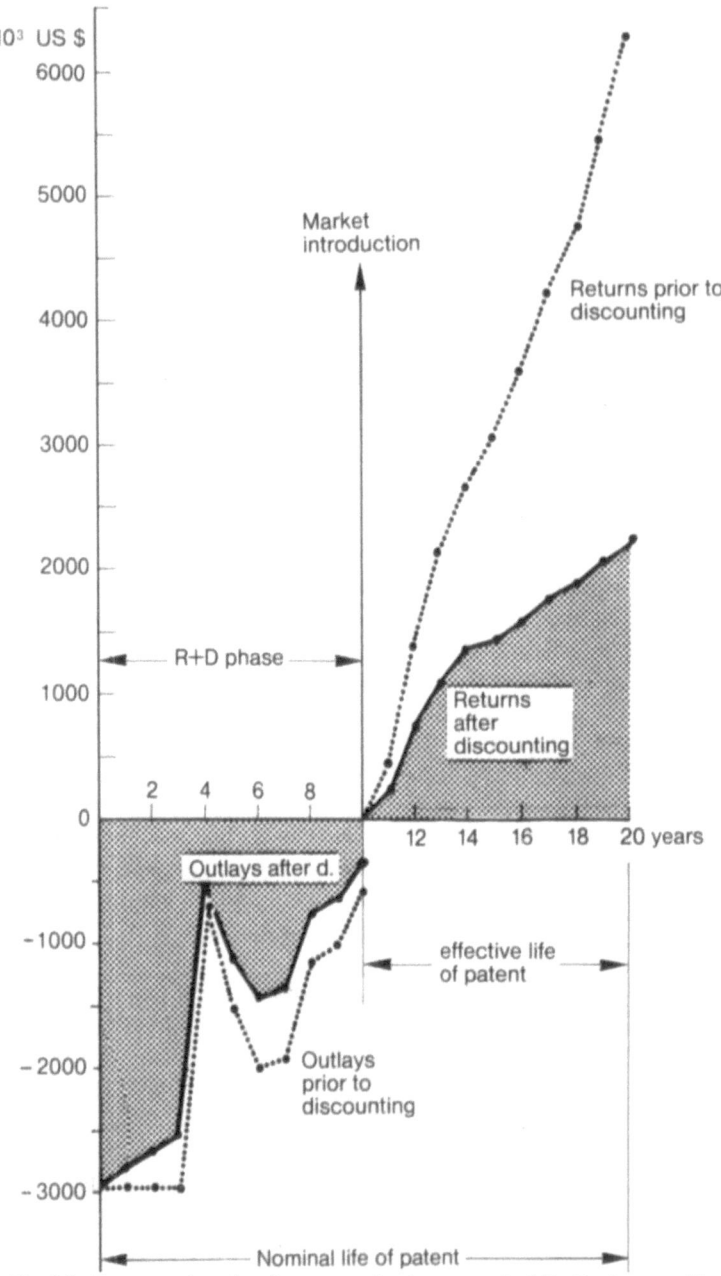

Fig. 5.2. Present value of an investment in pharmaceutical R+D. *Sources:* Hansen 1979; Virts and Westŏn 1981, Tables 3 and 9; author's calculations (1967 prices)

comparatively good to a comparatively bad investment?" This question amounts to the problem of finding that rate of discount (called the internal rate of return) which would reduce the present value of pharmaceutical innovation exactly to zero. Based on the cost data by Hansen [4] and the returns data by Virts and Weston [12] used in Fig. 5.2, one obtains an internal rate of return of 5.25% before tax. Recently, Joglekar and Paterson [5] have studied the fate of 280 new drugs introduced in the United States from 1962 to 1977 and have found a nominal aftertax *internal rate of return* of 10.7%. With an average rate of inflation of around 5% in the 1970s, this amounts to some 5.5% as a real value, which is not too far from the earlier estimate (which, however, does not take into account differences in tax treatment of financial assets and R+D investments).

Joglekar and Paterson [5] also illustrate the high degree of risk inherent in the pharmaceutical R+D process by estimating rates of return for the median rather than the average new drug. The problem with average figures is that they also include the very rare but extremely profitable "jackpot drugs," the most famous example being Valium of Roche. This average can be contrasted to the median, which is derived from asking, "Starting from the most profitable among those 280 new drugs, and stopping at No. 140 as the dividing line distinguishing the more successful from the less successful half, what is the value of the corresponding internal rate of return?" Because there are so few very successful innovations, the estimated median real rate of return is negative, −1.69% after a market life of 24 years [5, Table 1]. This underscores the riskiness of innovation, a characteristic shared by other industries, as illustrated by Mansfield [9] in a different context. In all, this leads to

---

Conclusion 6

The average real rate of return on pharmaceutical innovations introduced between 1967 and 1977 is in the neighborhood of 5%. For the majority of innovations, however, it lies below that value, the median rate of return reaching −1.7% after 24 years in the market.

It is interesting to note that the aftertax real rate of return on a corporate bond is around 2.3% on average [5, Table 1].

Looking at pharmaceutical R+D as an investment also makes it clear that extending the patent life of a drug is not the best means of encouraging innovation. True enough, extended patent protection will result in a continuation of the steeply rising return curve of Fig. 5.2, but due to discounting, profits made so many years after the original investment decision will not leave a big impact. Increased patent protection does create an advantage for firms that presently are cashing in on innovations launched a few years ago. But when it comes to a balancing of costs and returns associated with *future* innovations, prolonged patent protection will not make too much of a difference. The best solution would rather be a shortening of the registration process (with the Food and Drug Administration in the United States). In terms of Fig. 5.2, this cost saving will be likely to occur around year 8, resulting in a sensible shrinking of the "cost area" there.

Finally, it should be stressed that today market success may easily turn into financial failure. All it takes is a court ruling in favor of patients suffering from adverse drug

Table 5.1. Social and private rates of return on investment in innovations [8, p. 157 for part A; 13]

Innovation	Rate of return (%) Private	Social	Social, corrected for insurance, given coinsurance r^a: r = 50%	r = 25%	r = 10%
(A) *Industry in general*					
Primary metals innovation	18	17			
Machine tool innovation	35	83			
Components for control system	7	29			
Construction material	9	96			
Drilling material	16	54			
Drafting innovation	47	92			
Paper innovation	42	82			
Thread innovation	27	307			
Door control innovation	37	27			
New electronic device	Negative	Negative			
Chemical product innovation	9	71			
Chemical process innovation	25	32			
Major chemical process	31	56			
Chemical process innovation	4	13			
Household cleaning device	214	209			
Stain remover	4	116			
Dishwashing liquid	46	45			
Median	25	56			
(B) *Pharmaceutical industry*					
Pharmaceutical innovation X	40	65	53	46	43
Pharmaceutical innovation Y	74	160	117	96	85
Pharmaceutical innovation Z	19	69	44	32	24

[a]. The corrected rates of return for innovation X, for example, are calculated as follows. The private component of 40% remains unchanged. The unappropriated component has to be halved if coinsurance is 50%. This gives 40 + ½ (25) = 52.5%

reactions and presenting liability claims. The case of DES, an estrogen preparation that was launched in 1947, may illustrate the point. Between 0.5 and 2 million pregnant women who had used that drug gave birth to between 100000 and 160000 daughters from 1960 to 1971. Among these daughters, certain forms of cancer have turned out to be abnormally frequent. Some 40 years after market launch, 3800 women have filed claims against producers of DES totalling hundreds of millions of U.S. dollars [11]. Thus, an innovation that looks like an economic success may wind up a financial disaster.

The nature of the pharmaceutical R+D process also serves to qualify the high returns on capital of pharmaceutical companies that are often reported and criticized in the press. The above description of pharmaceutical R+D makes it clear that much of an innovator's capital is bound up in know-how that does not show up in physical assets. Moreover, it is good bookkeeping practice to exclude know-how from assets if it runs a considerable risk of not being incorporated in a marketable product. But this means that the capital base necessary to generate the profits is strongly understated in the books, causing profitability of the industry to look too favorable [see 3 for more details]. Summing up, we have

---

Conclusion 7
Compared with other investment opportunities of firms, pharmaceutical innovation is characterized by a high risk of failure during the R+D phase and, more
recently, even during the market phase. Return on capital as usually calculated
from the books therefore tends to be overstated.

Social Rates of Return on Pharmaceutical Innovation

In the case of well-informed consumers, their willingness to pay as a group can be used to gauge the social utility of a product, both existing and new. The condition of sufficient information is important because as can be seen from Fig. 5.1, a buyer is willing to pay a high price for a product having a high content of desired characteristics. Willingness to pay becomes a bad guide to actual social utility, however, if buyers typically are unable to identify the product's characteristics correctly.

For simplicity, it is assumed in the sequel that a patient has been sufficiently informed by the prescribing physician about the characteristics of the new drug in question. But with product characteristics given and known, an increased price of the drug makes it less attractive compared with competing brands. For example, let the breakthrough innovation B become so expensive as to exhaust the drug budget with just two rather than three units. In that event, the patient would have to choose between points 2B and 4T in Fig. 5.1, and given his indifference curves, he would prefer the cheaper traditional product T over B.

This *negative* relation between price and demand reappears in a simplified fashion in Fig. 5.3. Here, specification of the product in terms of quality is taken as given, and demand relates to an entire group of patients. The lower the price of the new drug relative to its competitors, the greater the quantity that will be demanded by informed patients in the course of a year. Vice versa, if the price to be paid is as high as P^m, the new product will not find its buyers. The line DD of Fig. 5.3 illustrates this relationship. Assume that the innovator sets a price P^*; economic theory has much to say about how this price is determined by the firm. At that price, the innovator can sell Q^* units per year. Total willingness to pay for these Q^* units of the new drug is given by the area OP^mAQ^*. The reasoning for this is as follows. The first few patients would be willing to pay the maximum price P^m, followed by others whose demand for the new drug is less urgent. The last few units before Q^* are sold to patients whose willingness to pay corresponds to the price P^* actually set by the firm. Aggregating these different subgroups of patients yields the area OP^mAQ^* as a measure of their total willingness to pay.

Most of the time, the seller is unable to discriminate between different groups of patients according to their willingness to pay. Therefore, he has to settle for a single price P^*. Revenue earned is given by price times quantity, represented by the area

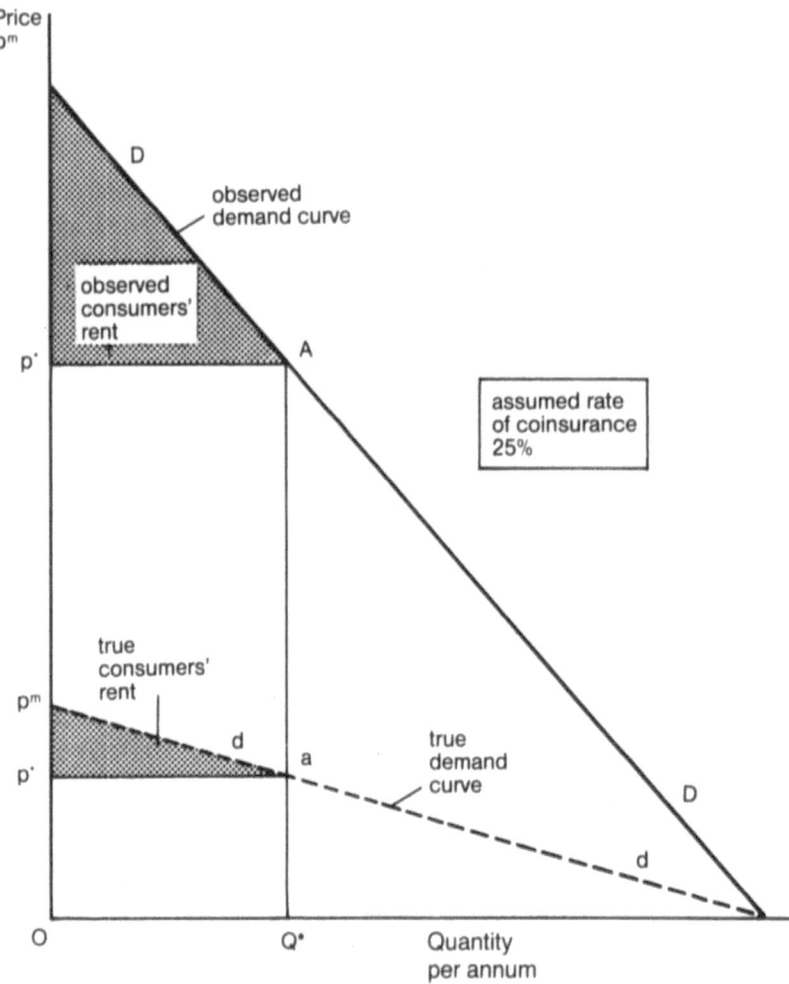

Fig. 5.3. Observed and true willingness to pay [15, p. 42]

OP*AQ* in Fig. 5.3. This sales revenue has to cover the cost of R+D, production, and marketing; not astonishingly, net utility of the new drug amounts to profit as the difference between revenue and these costs as far as the innovator is concerned. For the *buyers*, net utility is given by the difference between their actual willingness to pay and the amount they have to pay, i.e., the shaded triangle P*AP^m [8, pp. 154–156]. This argument can be given the form of

---

Conclusion 8
Because the innovator is unable entirely to appropriate patients' aggregate willingness to pay for the drug, his private return on the new product is only one component of the total social rate of return.

Table 5.1 taken from Mansfield et al. [8] and Wu [13], illustrates the cleavage between private and social rates of return to innovation. The first two columns suggest that in the case of pharmaceutical innovations, private as well as social rates of return may well be above average. But before jumping to the conclusion that pharmaceutical innovation is "better" than innovation in other industries, a problem peculiar to the pharmaceutical industry should be noted. This industry sells most of its products to buyers who are highly insured. From an economic point of view, insurance magnifies buyers' willingness to pay because it subsidizes their purchases.

For example, with cost sharing by the patient (the rate of coinsurance) set at 25%, the maximum acceptable price for the drug P^m is reduced to the net price p^m for the patient. This net price, which must be backed by actual willingness to pay, is only one-quarter of the price observed in the transactions. Or put another way, if the patient himself values the new drug at a maximum price p^m, the *observed* price at which the drug still can be sold to him is P^m, and this latter price is four times higher if coinsurance is 25%. Observed maximum price could be even higher if coinsurance were only 10%. More generally, it can be shown that aggregate utility of buyers as revealed by market transactions has to be deflated by the rate of coinsurance to arrive at the true value. In terms of Fig. 5.3, the area P^*P^mA is reduced to the area p^*p^ma, which is four times smaller. Therefore, social rates of return estimated on the basis of observed transactions can be rather misleading when transactions take place in an insured market. While the profit component of the social utility remains unchanged, the buyer component is overstated [14]. In Table 5.1, columns 3–5 show corrected estimates for three particular pharmaceutical innovations, using coinsurance rates of 50%, 25% and 10%. While social rates of return to these pharmaceutical innovations have to be adjusted downwards, they clearly remain comparable with those characterizing the remainder of industry. This gives rise to

---

Conclusion 9
Due to the influence of health insurance, social rates of return for pharmaceutical innovation cannot be based on observed market data without correction. However, even after correction, estimated social returns tend to remain of a magnitude comparable to those in other industries.

Implicitly, patients as a group have been taken as the arbiter for judging the social utility of pharmaceutical innovation. In view of the biases and uncertainties inherent in this procedure, other arbiters have been advocated, in particular insurers and Government agencies. But the question arises of whether these other arbiters are not biased themselves as well.

Insurers often decide whether they want to include a new drug in their list of benefits. However, even if these insurers are not for profit, they do not necessarily act in the best interest of their insureds and patients. A bias arises because most of the benefits of an improved treatment of rheumatism (say) do not accrue with the insurer. In the main, these benefits go to employers in the guise of a reduction of sick leave and indirectly to the remainder of the economy through an increase in the social product. On the other hand, the insurer has to cover the full cost of a new drug on his list, apart from coinsurance. Thus, the insurer as decision-maker will tend to exclude too many pharmaceutical innovations from covered benefits.

Another arbiter of pharmaceutical innovations might be the Government. On the one hand, government profits from increased social product due to a pharmaceutical innovation through increased tax revenue. This effect should make government officials more aware of the social utility of a new drug. On the other hand, many votes can be lost if a new drug causes severe adverse reactions. The particular decision-maker in an agency like the Food and Drug Administration may well see his job jeopardized when the media publicize deaths caused by a drug's side-effects. The many lives saved by the innovation largely go unnoticed, on the other hand [see 10 for an estimate]. Thus, slowing the rate of innovation may be a favored course of action as it also allows the costs of the policy to be shifted to a possibly different Government in the future. This argument comes down to

---

Conclusion 10
Admittedly, the social utility of a pharmaceutical innovation can be inferred only subject to many caveats. But health insurers and Government agencies can be shown to be biased too, owing to the influence of their self-interest.

Concluding Remarks

This paper started out from a curious observation: While technological change is heralded as the prime cost saver in the economy, it is feared by insurers and policy-makers in health care because of its cost-increasing effects. Indeed, there are forces at work in the health care sector favoring product innovation rather than cost-saving process innovation. The presence both of health insurance and of quality norms in the guise of requirements regulating marketing approval make efforts at reducing the production costs of an existing drug rather unattractive. However, a pharmaceutical firm will view the development of a new drug as one way of making a profit. Investment funds allocated to the R+D process necessary for pharmaceutical innovation have to compete with other investment alternatives, among them such a mundane option as putting the money in a bank.

When a cohort of would-be innovations is traced through time, most of them drop out during the R+D phase, which today stretches over no less than 10 years. And even those innovations which make it to the market may well prove a commercial flop. The average rate of return on investment in pharmaceutical R+D can be put at 5% at constant prices or at about 10% including an over time surcharge for inflation. This is just about comparable to the rate of return that can be earned on money in the bank.

Finally, it was noted that innovators can hardly ever reap the full benefits of their inventions. To the extent that they have to set a single price in a given market, many a buyer will pay an amount that is far lower than the actual price he would have been willing to pay. On the other hand, prices actually paid will often overstate true willingness to pay because of the leverage effect of health insurance. If an insured would go as far as to pay 5 dollars himself for a new drug, that drug may cost as much as 20 dollars if covered by a health insurance plan requiring 25% cost sharing. But even after correction for this bias, social rates of return on pharmaceutical innovation tend to be comparable to those in other industries.

Overall, pharmaceutical innovation, while using a good deal of an economy's resources in terms of highly educated manpower and physical plant, seems to be worth its price, at least on average. Its social payoff is as high as that of innovations in the remainder of industry, and it might well exceed that of many of the innovations taking place in the remainder of the health care sector.

References

1. Becker GS (1965) A theory of the allocation of time. Econ J 75:493–517
2. Chow GC (1967) Technological change and the demand for computers. Am Econ Rev 57:1117–1130
3. Comanor WS (1986) The political economy of the pharmaceutical industry. J Econ Lit 24:1178–1217
4. Hansen RW (1979) The pharmaceutical development process: estimate of current development costs and times and the effects of regulatory changes. In: Chien RI (ed) Issues in pharmaceutical economics. Lexington Books, Lexington, Mass., pp 151–187
5. Joglekar P, Paterson ML (1986) A closer look at the returns and risks of pharmaceutical R+D. J Health Econ 5:107–193
6. Lancaster K (1966) A new approach to consumer theory. J Pol Econ 74:132–157
7. Lancaster K (1971) Consumer demand: a new approach. Columbia University Press, New York
8. Mansfield ER et al. (1977) The production and application of new industrial technology. Norton, New York
9. Mansfield ER (1980) Comment on Zvi Griliches on returns to research and development expenditures. In: Kendrick JW, Vaccara BN (eds) New developments in productivity measurement and analysis. University of Chicago Press, Chicago, pp 454–461
10. Peltzman S (1973) An evaluation of consumer protection regulation: the 1962 drug amendment, J Pol Econ 81:1049–1091
11. Sigma of Swiss Reinsurance (1985) No. 9. Insert on liability insurance.
12. Virts JR, Weston JF (1981) Expectations and the allocation of research and development resources. In: Helms RB (ed) Drugs and health. Enterprise Institute, Washington, D.C.
13. Wu SY (1984) Social and private returns derived from pharmaceutical innovations: some empirical findings. In: Lindgren B (ed) Pharmaceutical economics. Liber, Malmö, pp 217–254
14. Zweifel P (1984) Technological change in health care: Why are opinions so divided? Managerial and Decision Economics 5:177–182
15. Zweifel P, Pedroni G (1985) Innovation and imitation. A delicate balance for economic policy (Innovation und Imitation. Eine wirtschaftspolitische Gratwanderung). Studien zur Gesundheitsökonomie 7. Pharma Information, Basel

6. The Medical, Social, and Economic Implications of Disease

B. S. BLOOM

Social Implications of Disease

According to Sigismond Peller, between 20% and 25% of children born in Europe in the sixteenth and seventeenth centuries died during their first year of life [1]. By 1750, infant mortality in Sweden and Britain had declined to 16%. The great killers throughout these centuries were infections, including infant diarrhea, and poor nutrition. At the turn of the twentieth century, infant mortality was reduced to 8.5% in Sweden and 12.7% in Britain due mainly to public health advances and improved food supplies [2]. Currently infant mortality is 6.5/1000 in Sweden and 10.0/1000 in Britain and the United States. Most deaths during the first year of life are now caused by genetic and birth-related problems and prematurity.

Peller also found that even during earlier centuries, those with economic and other advantages, for example the royal houses of Europe, suffered substantially less than most [3]. Galton showed, however, that once a high level of social and economic status was reached, additional advantages (in this instance, prayer) conferred no further benefits [4].

By the twentieth century, the problems of widespread disease and death from inadequate nutrition, unclean water, improper sewage disposal, and insufficient income in Western, industrial countries were well on the way to being solved [5]. These were the prime social and economic issues related to health and disease of *Homo sapiens,* and probably also to our near and distant ancestors.

Medical care controls very well, at least in high income countries, the great infectious and communicable diseases that decimated previous generations. These diseases, for example tuberculosis and diarrheal disease, had their roots in the social and economic vagaries of their time (Table 6.1). However, by the turn of the twentieth century, heart disease, malignant neoplasms, and stroke were already beginning to take their toll. Medical care now is occupied mainly with treatment of chronic life-style, environmental, and occupational diseases. Pneumonia and influenza, the ninth leading cause of death in 1984 (Table 6.1), are the only infectious diseases of the current top ten leading causes of death remaining from those of 1900.

Table 6.1. Leading causes of death, United States, 1900 and 1984

Rank	1900	1984
1	Consumption (including tuberculosis)	Heart disease
2	Pneumonia	Cancer (unspecified)
3	Heart disease	Lung cancer
4	Diarrheal disease (including cholera, dysentery)	Accidents
5	Unknown causes	Cerebrovascular disease
6	Kidney disease	Breast cancer
7	Typhoid fever	Motor vehicle accidents
8	Cancer	Chronic obstructive lung disease
9	Old age	Pneumonia and influenza
10	Apoplexy	Suicide

The Primacy of Societal Cost

At the same time as we take for granted the broad population benefits of antibiotics, immunizations, psychotropic drugs, anesthesia, and aseptic surgery, government and the academic community complains of the cost of medical care. In fact, cost is the single most important issue of health policy in nearly every Western country; access, quality, equitability, and the like are considered minor concerns relative to the increasing amounts of societal resources devoted to medical care. Parenthetically, the general population seems little concerned with cost, and appears prepared to devote even more resources to medical care, as a recent opinion survey showed [6].

Researchers have been very successful in measuring societal costs of medical care, especially in Western countries. For example, in the United States unadjusted medical care costs increased threefold between 1950 and 1965, the year of enactment of Medicare (insurance that pays for acute care for the elderly) and Medicaid (payment for services for the poor), and 12-fold between 1965 and 1986 (Table 6.2). Between 1950 and 1986 per person expenditures increased more than 20-fold (Table 6.2). The proportion of gross national product going to medical care increased 250% between 1950 and 1986 (Fig. 6.1). An increasing percentage of the elderly among the population, greater intensity of services, new technology, and inflation are among the most important contributors to rising costs.

In modern times the hospital has been the largest consumer of medical resources. One-third of all medical expenditures went for hospital care in 1960 (Fig. 6.2). By 1986, the figure had risen to 42%. The percentages going to the physician and dentist have remained constant over the past two and one-half decades, while the amount spent on nursing home care, mostly for the elderly, has increased more than fourfold.

Table 6.2. Expenditure on medical care in the United States

	Total	Per person
1950	$ 13 Billion	$ 82
1965	$ 39 Billion	$ 198
1986	$ 465 Billion	$ 1820

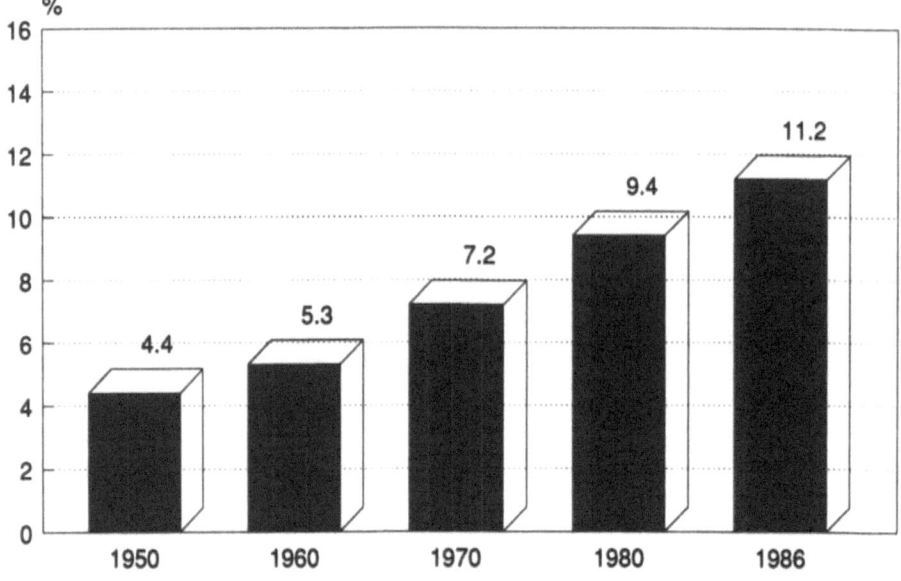

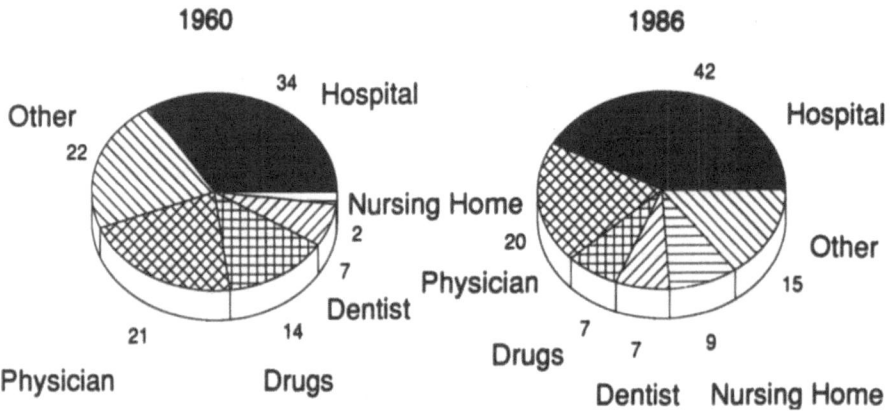

Fig. 6.2. Unites States' medical care expenditures, 1960 and 1986, in %

The increased funds devoted to hospital and nursing home care are being financed by reductions in the percentage going to pharmaceuticals and all other services. In fact, the portion of total medical care expenditures spent on drugs and medications in the United States was halved between 1960 and 1986.

The percentage of medical care expenditures paid by the U.S. government has increased steadily, nearly doubling over the past 26 years (Fig. 6.3). Medicare and Medicaid undoubtedly are the main cause of this increase. Private insurance in 1960

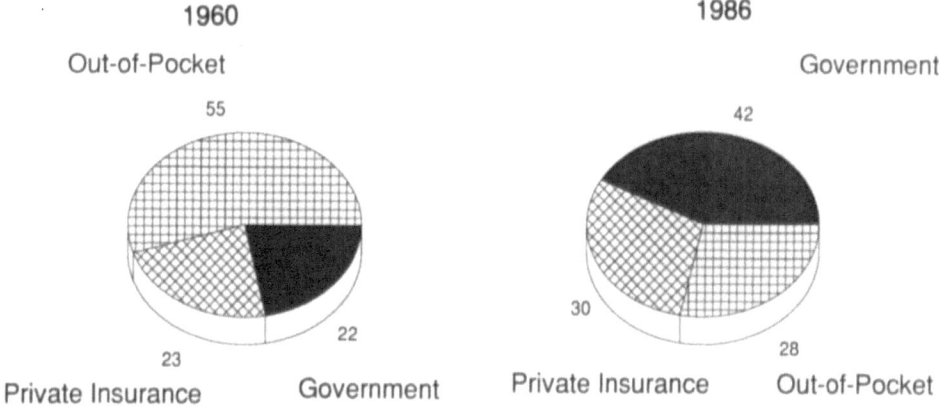

Fig. 6.3. Source of medical care funds in the United States, 1960 and 1986, in %

paid for 23% of all medical care and 30% in 1986. At the same time, the percent paid by patients out of their own pocket has been halved.

The Cost of Disease Treatment

While we know a great deal about societal and government expenditures for medical care, we know rather little about the social and economic costs of specific diseases for individual patients or specific subgroups of patients, and the distribution between direct medical and indirect costs. A recent study of system and family costs of caring for children with cancer is one exception [7]. For example, what are the direct medical costs of treating patients with chronic obstructive lung disease? What are the indirect costs such as early death, lost productive capacity, and early retirement on family, industry, and society? In essence, we know what, from whom, and to whom for all people, but not for any one person. The results of a selection of studies highlight the direct medical and indirect economic and social costs related to medical care.

Direct Medical Costs

The first example is a study of direct medical costs of treating 441 patients with angina pectoris. They were followed for up to 2 years from date of first entry to the study until death or study completion. All Washington, D. C., Medicaid recipients newly diagnosed for angina pectoris on or after 1 December 1981 were included in the study and were followed through 30 November 1983. All participants had to be Medicaid-eligible for at least 6 months prior to admission in order to be certain they had no other coronary heart disease. Patients with any preexisting history of cardiovascular disease, except essential hypertension, were excluded from the study.

All Medicaid payment claims for services were obtained from the Medicaid Management Information System computerized data files for each person identified as being eligible for the study. A full history was constructed of Medicaid-paid services utilized, and cost, for their heart disease. As all study participants were continuously eligible for Medicaid benefits, it is unlikely that any important medical care services were obtained outside the study.

Patient demographic information included age at study inception, sex, race, and aid category. As newly diagnosed patients were admitted and/or lost throughout the 2-year study period, all reported costs associated with the disease for each participant were adjusted by number of months of participation in the study. Therefore, reported costs represent mean expenditures by Medicaid per patient-month. All direct medical expenditures for care for these individuals were collected and the natural history of therapy and medical costs were determined.

The 441 people who met criteria for inclusion in the study accounted for 5181 person-months, or a mean of 11.7 months per person in the study. Nearly three-fourths of the study population were female, 51.4% were aged 65 and above, and 93.2% were black. The high percentages of blacks and females reflect the constituency of the Washington, D.C., Medicaid population.

Table 6.3 shows the total costs of care for the study population adjusted for the number of months each patient was in the study. The mean cost of caring for patients with angina pectoris was $ 62 per month. Approximately half went for hospital care and the remainder was equally divided between physician care and drugs. However, these data are skewed to the left by the very few patients who had an operation – cardiac catheterization and/or coronary artery bypass graft. Table 6.3 also shows the marked difference in cost between those who did and those who did not have an operation. A full analysis of the costs of these procedures is not possible as only four study participants had either of these procedures.

Table 6.3. Mean treatment cost ($) per person-month for patients with angina pectoris

	Patients Total cost (n = 5181)	Patients with Operation (n = 52)	Patients without Operation (n = 5129)
Hospital	32	1195	21
Physician	15	292	12
Drugs	15	17	15
	62	1504	48

Mean monthly costs, however, do not show the great variation in cost facing patients over time (Fig. 6.4). Costs varied in fairly uniform 4-month cycles. As would be expected, costs were high when the disease was first diagnosed, after which they declined. There was an increase in monthly costs at approximately 4 months and 8 months after initial diagnosis because of physician revisits, hospitalizations, and changes in drug therapy. Costs then decreased sharply and remained stable at a low level thereafter until the study ended.

Twenty-nine (80.6%) of the 36 hospitalizations occurred during the first 8 months of therapy. Additionally, operations were performed in months 3, 4, 6, and 8,

Cost in $

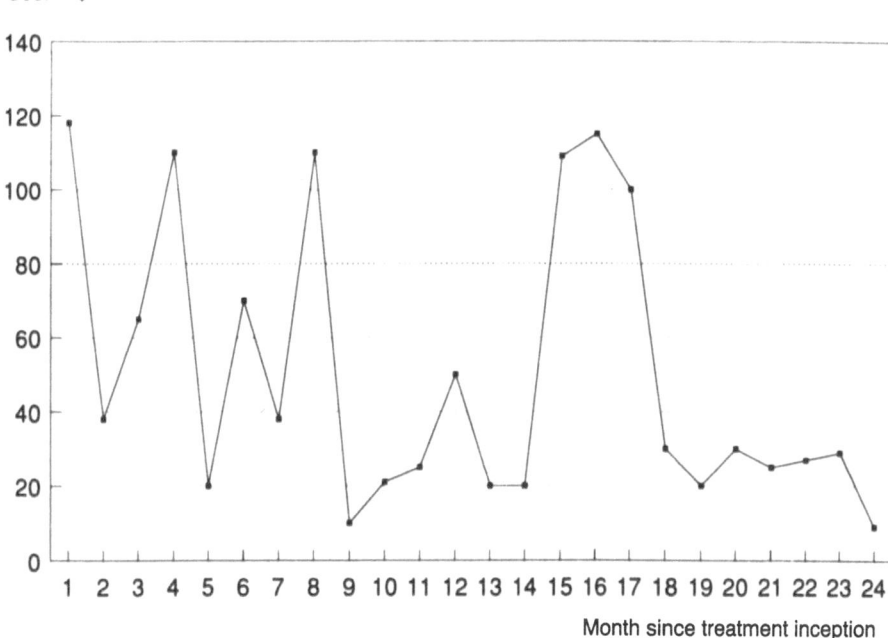

Month since treatment inception

Fig. 6.4. Mean per patient-month cost by treatment month

explaining large variances in costs for these months. Variations in physician costs showed no real trends, while drug costs were quite stable during all months, following the initial high cost of the first month.

A straight line was fitted to the total cost data by simple linear regression. The y-intercept ($ cost) indicates the predicted y-value of the regression line when x (time) equals zero. The results in this instance were:

y-intercept   = $ 88
slope   = −10.8

Broadening the Definition of Direct Medical Costs

In the few instances when disease-specific costs of therapy are examined, effects of exogenous and endogenous factors such as side-effects of therapy are often over-looked. We are all well aware that every therapy carries a risk of side-effects, from the simple such as taking an aspirin to the complicated such as liver transplant. These risks are usually low, although they can be catastrophic in rare instances [8].

The second study determined the cost of caring for people with arthritis. Also included was the cost of adverse drug reactions. It is well accepted that many patients on nonsteroidal anti-inflammatory drugs have gastrointestinal side-effects sufficient to warrant additional medical care. There is also a direct relationship between compliance with nonsteroidal anti-inflammatory drugs, control of symptoms and control of disease, and gastrointestinal side-effects. The problem, of course, is that pain relief is obtained at a lower drug regimen and dose than anti-inflammatory response. Patients are often satisfied with the former alone while physicians try to achieve both. Thus compliance becomes a major issue. Therefore, it is necessary to determine compliance with the drug, the direct medical cost of arthritis treatment, and the direct medical cost of treating the gastrointestinal side-effects in order to gain a good understanding of arthritis treatment.

Patient cooperation with the drug regimen prescribed by the physician was determined by retrospective examination of pharmacy refills, after the method of Deyo et al. [9] and based on data from the Washington, D.C., Medicaid program. The rate of compliance was calculated by dividing the number of days supply, as indicated by the number of tablets, capsules, or pills provided by the pharmacy, by the observed number of days between the date the prescription was filled and the date the prescription was refilled. If the physician prescribed a 30-day supply and the patient refilled the prescription after 40 days, the compliance rate was calculated as $30 \div 40$, or 0.75. If, on the other hand, the prescription was refilled after 28 days, the compliance rate would be 1.07 ($30 \div 28$); all rates greater than 1.00 were defined as equal to 1.00.

Refilled prescriptions were stratified by the following dosing schedule:

q.d. – up to one pill, tablet, or capsule per day
b.i.d. – 1.01 to 2 per day
t.i.d. – 2.01 to 3 per day
q.i.d. – 3.01 to 4 per day
more than q.i.d. – 4.01 and above per day.

A series of paired comparisons were performed among dosage schedule groups. Statistical significance was determined by t-test. The Bonferroni method was used to overcome distorted significance level problems of multiple comparisons [10]. The methods of Cook and Campbell were employed to deal with limitations of adjustments using analysis of covariance with nonequivalent groups [11]. Only patients who had a rate of cooperation with the standard and approved anti-inflammatory dosage regimen of at least 0.40 were included in the study. All costs of care provided to patients for their arthritis during each 3-month period were aggregated. Costs were adjusted by specific arthritis diagnosis and compliance rate.

A gastrointestinal adverse drug reaction was defined as any care provided by physician, outpatient clinic, inpatient hospital, or outpatient pharmaceutical service with a diagnosis of peptic ulcer, gastritis/duodenitis, other disorders of the stomach or duodenum, and/or gastrointestinal symptoms, e.g., dyspepsia, heartburn, or a pharmacy claim for H_2 blocker, sucralfate, or antacid, which occurred during the antiarthritis treatment study period. Patients could not have any such gastrointestinal diagnoses or pharmacy claim during the preceding 12 months. Treatment costs and gastrointestinal side-effects were summed to obtain the total direct cost of care related to arthritis.

The rate of patient compliance, adjusted for patient age, sex, and race, was highly correlated with the number of tablets, capsules, or pills the patient was to take, as prescribed by the physician (Fig. 6.5). Patient compliance was related to demographics, i.e., older patients were generally more compliant than younger ones, males

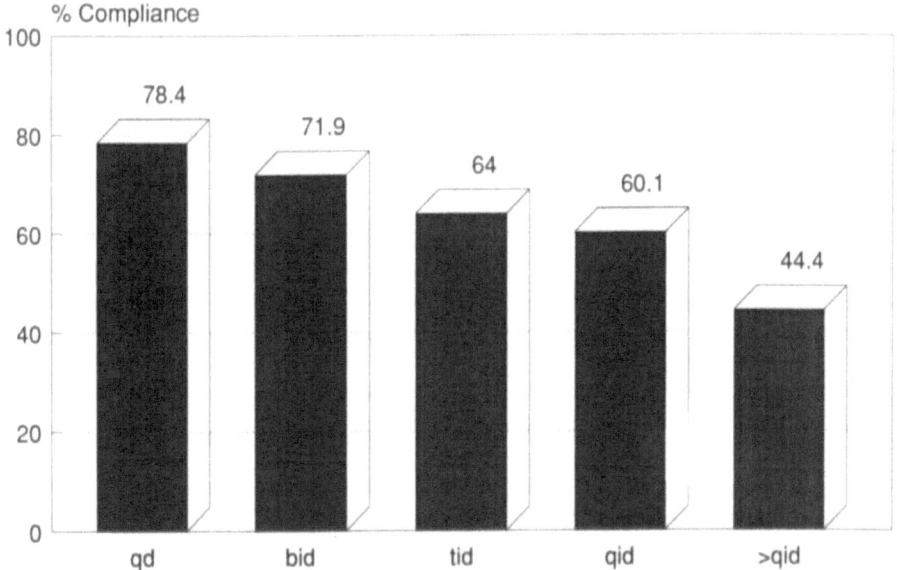

Fig. 6.5. Adjusted mean compliance with NSAIDs by dosing schedule

more so than females, and whites more so than nonwhites. For patients on a once-a-day regimen, the mean compliance rate was 0.78, while the rate was 0.44 for those whose regimens necessitated doses more than four times per day.

Table 6.4 presents the direct medical costs of care for treating patients with arthritis. The distribution of expenditure data showed many observations with zero or very low values and only a few observations with very high expenditures, usually the cost of hospitalizations. The costs of treating arthritis were, to a large extent, the cost of the NSAID used. Pharmaceutical claims per quarter averaged $78 or 53.8% of total arthritis treatment cost of $145. The remainder was nearly equally divided among inpatient hospitalization cost and physician or outpatient clinic fees.

Any discussion of the cost of arthritis treatment must also include the cost of iatrogenic disease caused by the therapy. We found that across all treatment regimens approximately 25% of the study population had a gastrointestinal side-effect, assumed to be related to the NSAID, sufficient for the patient to seek additional

Table 6.4. Mean treatment cost ($) of arthritis treatment, per quarter

	Arthritis treatment	GI adverse drug reaction	Total
Hospital	31	25	56
Physician/Clinic	36	13	49
Pharmaceutical	78	28	106
	145	66	210

medical care. In the vast majority of the instances, an additional drug, usually H_2 blockers although occasionally sucralfate or an antacid, was prescribed. In rare instances, a patient was hospitalized. Pharmaceutical claims accounted for 42.4% ($ 28) of total adverse drug reaction treatment cost ($ 66) while the few hospital claims accounted for 37.9% and physician claims for 19.7% of total expenditures for gastrointestinal side-effects.

Therefore, arthritis treatment costs for this study, including the cost of treating the disease plus the cost of treating adverse drug reactions, totaled $ 210 per 3-month period (Table 6.4). Arthritis treatment costs were 69.0% while the cost of gastrointestinal adverse drug reactions was 31.0% of total costs. The cost of adverse drug reactions added 45.5% to the cost of arthritis treatment.

Indirect Costs of Policies Designed to Save Money

The third study examined perverse effects of government action to control medical expenditures. In 1982 one state in the United States, in an effort to reduce its expenditures for medical care for the poor, severely restricted access by reducing the number of medications on their approved formulary. In many instances it was difficult to determine the rationale for removing specific drugs from the approved formulary, particularly those previously proven cost-effective. For one such drug, cimetidine, we measured the effects before and after the imposition of this policy [12].

Diagnostic, demographic, and health care utilization data for each study participant were obtained from West Virginia Medicaid. All patients, and the payment for services provided, had to include a diagnosis of peptic ulcer. The analysis covered two study periods, each of which lasted 9 months. In the first time period, preceding the restricted formulary, any drug could be prescribed (open formulary period), while during the latter study period the number of drugs was restricted (closed formulary period). A 3-month period between open and closed formulary was allowed to elapse for adjustment by patients and physicians. Equivalent periods were examined in consecutive years to avoid seasonal bias.

Results showed that the total Medicaid cost of treating patients diagnosed with peptic ulcer declined by 15.0% between open and closed formulary periods. This was due entirely to a 21.2% reduction in the number of patients treated. Adjusted treatment cost per patient increased 23.6% for hospital care and 3.1% for physician services while it declined 83.6% for pharmaceuticals.

However, we estimated that short-term savings made during the 9-month closed formulary period by excluding a proven cost-effective drug for the treatment of peptic ulcer would soon be overwhelmed by the sharp long-term increase in expensive inpatient therapy for those few patients needing such medical care. In essence, the savings accruing from the alternative to a low-cost drug for the many would be outweighed by high-cost inpatient care, including operations, for the few. In support of this hypothesis, during the 9-month closed formulary period the rate of diagnostic endoscopies increased by 30.6% while the rate of surgical operations for peptic ulcer disease increased by 52.8%, as compared with the rates in the open formulary period (Table 6.5).

Table 6.5. Surgical procedures for peptic ulcer, by formulary status; rate/1000 patients

	Open formulary	Closed formulary	% change
Diagnostic endoscopy	23.5	30.7	+ 30.6
Surgical procedures	22.9	35.0	+ 52.8

Indirect Social Effects

The last study investigated the broad social and economic effects of disease treatment. When such effects are measured they often show costs or benefits greater than those of direct medical care. However, we rarely document these effects over time as disease changes or as external factors influence the indirect effects of disease and therapy on defined populations.

Results of multinational epidemiological studies in Belgium, the Federal Republic of Germany, Sweden, the United Kingdom, and the United States found sharp declines in work loss for persons with peptic ulcer and gastritis/duodenitis. Data from England and Wales and the Federal Republic of Germany show this clear change (Fig. 6.6). The trend was stable or declining slightly prior to the introduction of a new pharmaceutical therapy. After wide use of the drug, work loss declined steadily. Monetarization of days of work loss avoided would undoubtedly show very large savings.

Analysis of United States data on changes in rates of early retirement due to disability caused by peptic ulcer disease and related disease showed a sharp decline in

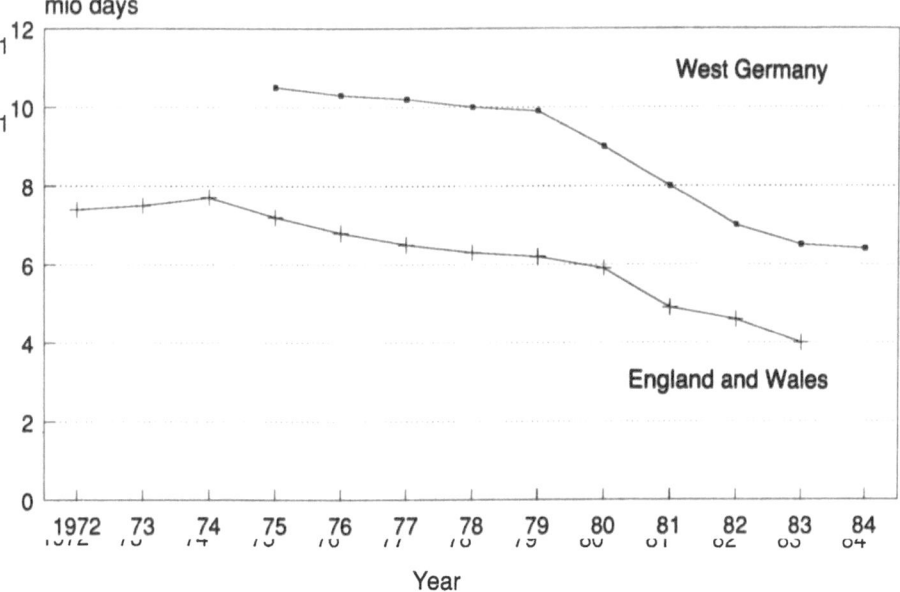

Fig. 6.6. Days of work loss due to peptic ulcer and related diseases

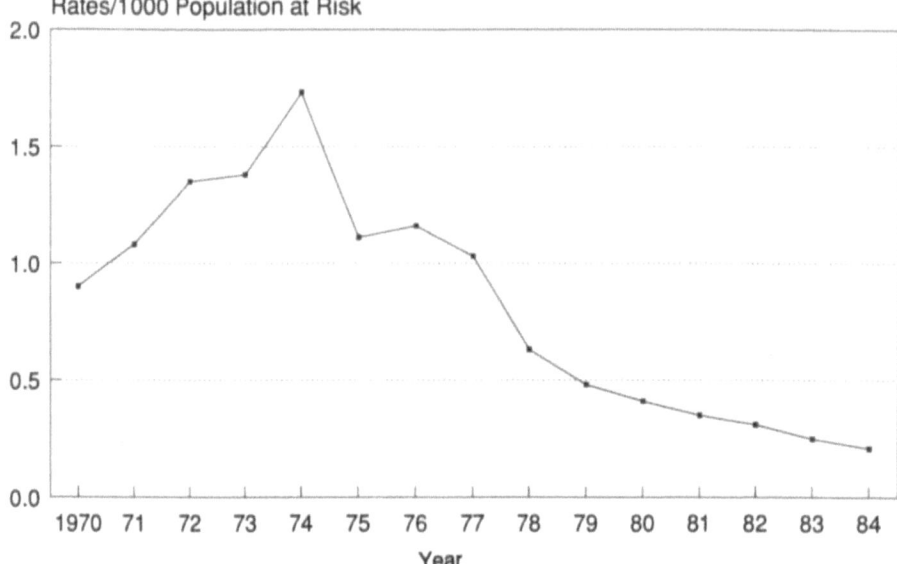

Fig. 6.7. Early retirement due to peptic ulcer and related diseases

the rate/1000 of the United States population at risk after the introduction of histamine H_2 receptor antagonists (Fig. 6.7). The rates of early retirement were increasing before the drug was available and showed a sharp decrease thereafter, in spite of increasing rates of early retirement for most other diagnoses. Here, too, placing a monetary value on early retirements averted would increase societal benefits greatly. Reductions in work loss and early retirement, however, do not quantify the symptoms and other adverse effects individual persons do not undergo.

In conclusion, we can only know the extent and size of medical, social, and economic effects of medical care if we measure them. We have tools and techniques to measure efficacy, effectiveness, cost, quality of life, and other direct and indirect effects of medical care – prevention, treatment, and rehabilitation. Some methods or instruments may be at a more advanced technical state than others, but increasing adaptation and use will improve their sensitivity, specificity, reliability, and validity. There are no longer any good reasons not to measure.

References

1. Peller S (1943) Studies on mortality since the Renaissance. Bull Hist Med 13:429–461
2. Bloom BS (1984) Changing infant mortality: the need to spend more while getting less. Pediatrics 73:862–866
3. Peller S (1965) Births and deaths among Europe's ruling families since 1500. In: Glass DC, Eversley DEC (eds) Population in history: essays in historical demography. Edward Arnold, London, pp 87–100

4. Galton F (1872) Statistical inquiries into the efficacy of prayer. Fortnightly Rev 68:125–135
5. McKinlay JB, McKinlay SM (1977) The questionable contribution of medical measures to the decline of mortality in the United States in the twentieth century. Milbank Mem Fund Q 405–428
6. Blendon RJ, Altman DE (1984) Public attitudes about health care costs. N Engl J Med 311:613–616
7. Bloom BS, Knorr RS, Evans AE (1985) The epidemiology of disease expenses: the costs of caring for children with cancer. JAMA 253:2393–2397
8. Carson JL, Strom BL, Soper KA et al. (1987) The association of nonsteroidal anti-inflammatory drugs with upper gastrointestinal tract bleeding. Arch Intern Med 147:85–88
9. Deyo RA, Inui TS, Sullivan B (1981) Noncompliance with arthritis drugs: magnitudes, correlation and clinical implications. J Rheumatol 8:931–936
10. Snedecor GS, Cochran WG (1980) Statistical methods. Iowa State University Press, Ames, Iowa
11. Cook TD, Campbell DT (1979) Quasi-experimentation, design and analysis issues for field settings. Rand McNally College Publishing Co., Chicago
12. Bloom BS, Jacobs J (1985) Cost effects of restricting cost-effective therapy. Med Care 23:872–880

7. From Medical to Socioeconomic Evaluation of Drug Therapy

J. L. READ

Despite the growing interest in socioeconomic evaluation evidenced by this symposium, there is still a considerable amount of ignorance and lack of acceptance among an important group of decision-makers: the physicians who write the medical literature and who interpret it for their colleagues.

For the purposes of this paper, the group in question will be called "clinical policymakers." This includes those doctors who perform clinical trials to evaluate drugs, surgery, and other forms of therapy. Investigators reporting on the clinical value of diagnostic procedures are also included, as are most epidemiologists. Whenever these researchers report results along with recommendations that a particular clinical intervention is efficacious, dangerous, or to be preferred over another, they are formulating *clinical policy*. Naturally, those who write editorials in medical journals are clinical policymakers. So are doctors who may not write articles, but review the literature and present their opinions in Grand Rounds and clinical conferences at every hospital in the world.

It may be useful to pause at this point in the development of socioeconomic evaluation of drugs to consider where most clinical policymakers now stand in terms of understanding these new measures. I shall explore the reasons for what I perceive as a lingering sense of distrust, or at least, disinterest on the part of these key decision-makers. Then, I shall consider some steps which might be taken by those working in this young field to improve the understanding of their efforts.

Socioeconomic evaluations are concerned with the impact of a health care intervention on patients' quality of life and on the economic costs or savings to different parties. While some medical editorialists are beginning to argue for the importance of demonstrating an intervention's impact on quality of life and cost-effectiveness, regular readers of the medical literature – that is, the large circulation, peer reviewed, general and subspecialty journals – are aware that end points such as these rarely appear in research articles. Furthermore, data bordering on these areas from patient questionnaires are often disparaged as unreliable or "soft," somehow less useful than those derived from "hard" clinical measures.

Lack of Acceptance Should Not Be Surprising

Why are socioeconomic evaluations so foreign to most clinical policymakers? At least four possibilities are worth considering:
1. These measures constitute a new language for doctors.
2. They are based on conceptual models which diverge from the way doctors traditionally view disease.
3. They are a response to changes in the basic rules by which doctors work.
4. They address the information needs of a changing cast of decision-makers.

A New Language for Doctors

The lexicons of quality of life assessment and cost-effectiveness analysis are far removed from the vocabulary which doctors use to describe clinical outcomes. Data from traditional clinical evaluations follow the lines in which doctors think about data gathering for patient care: verbal reports of symptoms in the patient history (anamnesis), findings from the physical examination, and then the results of laboratory tests.

Study results are thus reported in terms of "percent reporting improvement," "number of adverse reactions," "documented cases of finding X," and, most emphatically, in terms of the results of technological tests. Test results, whether from the "wet laboratories," the X-ray suite, or the myriad of physiological laboratories in a modern hospital are virtually a requisite for publication. They are easily quantifiable and seem to lend themselves more readily to statistical examination.

In contrast, the arcane scaling systems of health status indices are bewildering to clinicians because they do not flow naturally from the traditional patient history, physical examination, or laboratory findings. Further transformations of outcome variables into quality adjusted life years (QALYs) are even less familiar. And economic end points are, up to now, completely outside their scope of interest. Discounted present value, willingness to pay, and human capital valuations are still as foreign (and disturbing) to most clinical policymakers as are autopsy findings to most economists.

Different Models of the Problem of Interest

When we consider a decision, we hold a conceptual model of the problem in our minds (although sometimes, this model is less explicitly formulated than others). For example, a doctor managing a patient with angina pectoris might be working on the basis of a model such as that shown in Fig. 7.1. In this medical model of cardiovascular disease, there are causal links between environmental and hereditary factors, microscopic pathological processes, physiological consequences, and symptomatic manifestations. Intervention can change the likelihood of future symptoms and events such as myocardial infarction or death.

Of course, doctors do not review all of the elements of this model each time they see a patient. Instead, they tend to follow "clinical policies" or heuristics (decision rules)

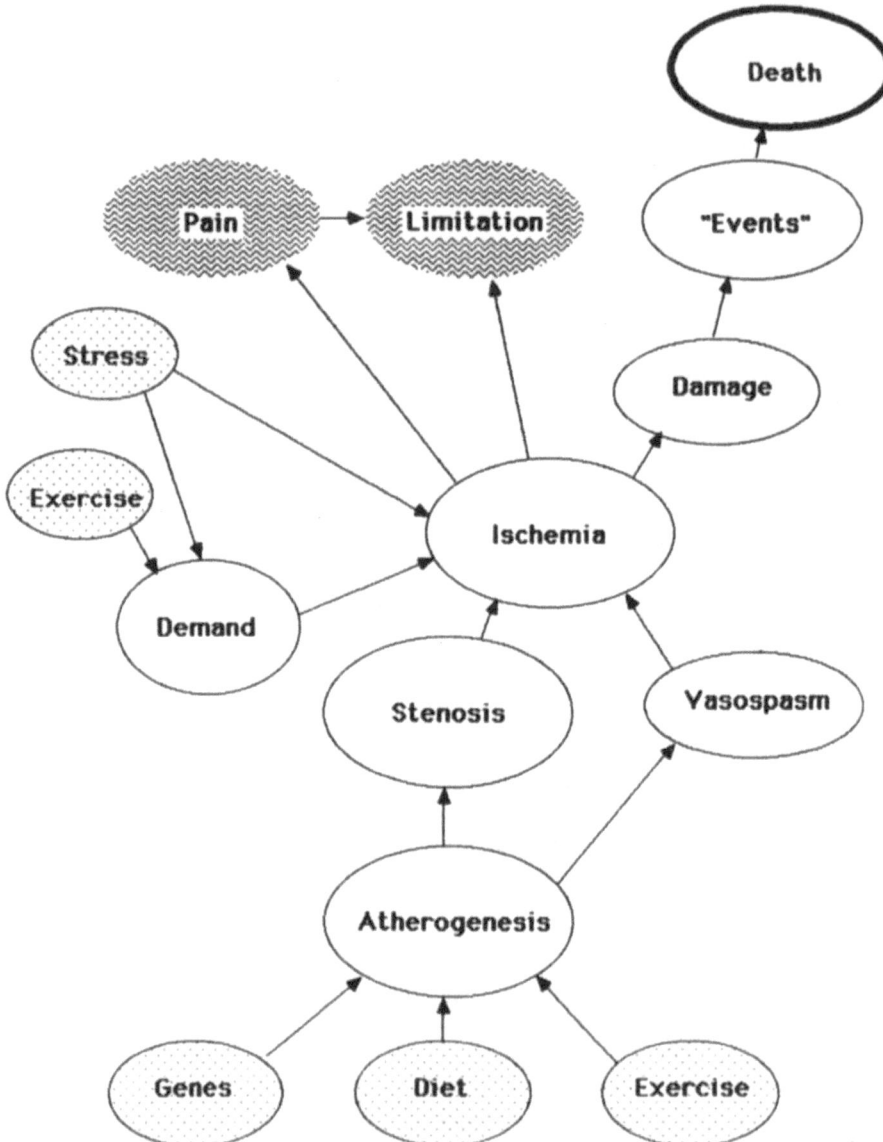

Fig. 7.1. A medical model of cardiovascular disease

which dictate that a particular test, drug, or other intervention is appropriate in a particular setting. These clinical policies are changed only when new knowledge causes the doctor to revise the details of the underlying conceptual model.

Models such as these are also brought into explicit play in the design of and choice of outcome measures for a clinical trial. Medical science has developed ways to measure all of the elements which appear in these disease models. In an important way, the availability of acceptable assays or probes actually defines the nature of the model. There is a strong correspondence between the appearance of an element in the model and the availability of a test to measure that element.

Medical thinking is fuzzy, however, about the elements having to do with patient symptoms and related concepts. Is it that the tests for these elements are less well developed? Do doctors not trust their own ability to measure these types of outcome? Perhaps it is because the assessment of patient feelings, satisfaction, and quality of life have been relegated to the "art" rather than to the "science" of medicine.

Social scientists, on the other hand, have their own models of the world, as do economists. Figure 7.2 shows a simple example of how the world might be viewed by

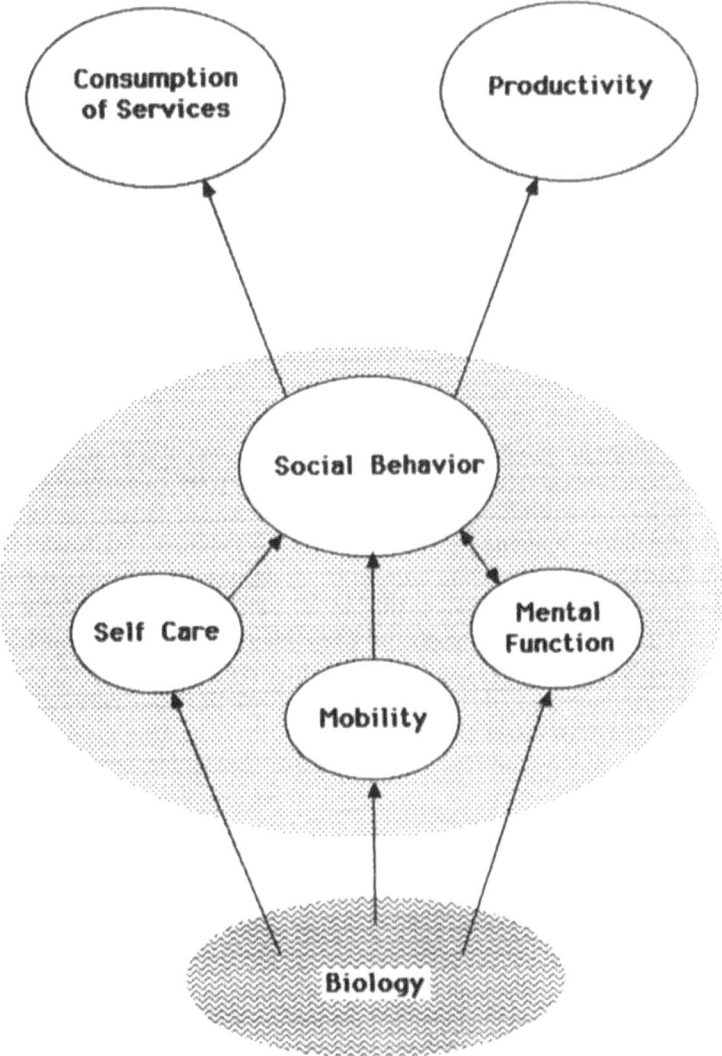

Fig. 7.2. A socioeconomic model

the hyphenated mind of a practicing "socioeconomist." Here, the measures of health status and economic outcomes are more or less well developed, while the underlying biology is the fuzzy zone.

It is possible to see how doctors, operating from their medical models, and those advocating socioeconomic evaluation of health care interventions would easily talk past each other because there is not enough common territory in their respective models of the world.

The Rules of the Game Have Changed

Over the course of the twentieth century, the basic tenets by which physicians regard their actions have undergone dramatic changes. In the preantibiotic era, the watchword was *primum nil nocere – First Do No Harm*. This was just as well, since medical science had relatively little to offer most patients. It was important that the nostrums and elixirs were, at lest, harmless.

With the great advances in diagnostic techniques and drug therapy occurring in midcentury, medicine could finally offer genuine relief to many patients. Doctors soon found themselves fighting a losing battle to keep up with the pace of scientific advance. By the 1960s, the watchword had changed from Do No Harm to *Be Complete*. Thoroughness became the highest professional virtue. Medical students and house officers were trained to be obsessed with the medical chart, which had to be "buffed and tuned" overnight for teaching rounds the following morning. Out of the respect earned by the laboratory sciences for very real advances grew an almost religious obeisance to the "technological imperative" – a reliance on technical procedures rather than touch and feel.

Then, sometime in the last decade or two, the rules changed again. Those who manage society's investment in the health care enterprise discovered that doctors could do more than was affordable. (Operating under the old rule, doctors actually seemed compelled to do more.) Thus, a new rule was imposed, this time from outside the medical profession: *Give Value for Money*. Physician test-ordering and prescribing behavior is now under intense scrutiny by organizations which must control health care costs. Trainees are being taught to think about cost and likely benefit before ordering a test or prescribing a drug. Because this new mandate often runs afoul of a physician's responsibility to do as much as possible for the individual patient at hand, it inevitably generates resistance. That resistance also extends to the types of socioeconomic evaluation which support value-for-money decisions.

New Decision-makers

Health care delivery is evolving from a cottage industry of individual practitioners following their own lights to a highly organized enterprise of large-scale proportions. Medicine in the United States is undergoing major structural changes, moving toward greater centralization of planning and control. Even in the absence of national health insurance, American medical practice is growing more and more like the government-operated systems in other Western countries, in that doctors must share decision-making power with managers who are concerned with optimal allocation of limited budgets. European systems which have long had centralized financing are tightening controls and strengthening incentives which affect individual practitioners.

As well described elsewhere in this symposium, regulatory and financing authorities now ask that the net cost impact of new therapies be evaluated before they are approved for marketing or reimbursement. What could be farther from the contemporary physician's role than to preside over a fixed budget, doling out scarce resources in an attempt to buy as much health for the dollar as possible? While this sounds more like the wartime rationing of penicillin or battlefield triage, it is precisely the task faced by contemporary health care administrators.

Increasing Appreciation of Socioeconomic Evaluations

Some researchers have done a commendable job of making quality of life assessments and cost-effectiveness analyses comprehensible to clinical policymakers. There remain several key opportunities, however, which deserve more attention from those who wish to advance the influence of socioeconomic evaluations. These include better demonstration of the linkages between models, cross-fertilization of measurement theories, empirical correlation of alternative measures, demonstration studies, and expanding physician opportunities in new decision-making roles. Each is briefly discussed below.

Link the Old and New Models

The first step in knitting conceptual systems together is to make the socioeconomic model more explicit. Reports of studies with quality of life end points should make it clear which components of the multidimensional construct of health are meant to be captured. Until standardized measures become familiar to the medical reader, it is not sufficient to reference the papers which document an index's development. Instead, the methods or discussion sections of the report should be used to summarize the aims of the health status instrument, its content, and data supporting validity and reliability.

The next step is to explain the investigator's view of why this particular measure is a relevant end point for the intervention in question. This model should be presented as it was held by the investigator a priori, before the data came in, as well as the rationale which emerges in exploring correlations among end points.

A cost-effectiveness analyst has a special obligation to present explicit models. There must be no doubt in the reader's mind as to which categories of financial effects and health effects are included in the analysis. The perspective from which the costs and savings are identified must be made clear, preferably by describing the types of decision-makers who hold those perspectives (e. g., Director of an HMO, a Minister of Health, of chief of a Hospital Pharmacy). Costs, savings, and health benefits and risks are often distributed unevenly over different payors and patients. Assumptions regarding these distributive effects should be explained as transparently as possible.

Of course, the choice of the *effectiveness measure* is critical to the persuasiveness of these studies.

Cross-fertilize Measurement Theories

The approaches by which doctors decide whether or not a test is useful could benefit from some of the thinking which goes on in the social sciences. The uses of medical tests fall into roughly three categories:
1. To classify patients as having disease or not (diagnosis)
2. To follow the progress of disease and therapy (monitoring)
3. To compare groups of patients (as in clinical trials)

   In recent years, considerable advance has taken place in the way doctors think about diagnostic testing. The qualitative concepts of sensitivity and specificity are widely understood and there are now people in every medical center who can, if interested, calculate the counter-intuitive, Bayesean, predictive value positive. On the other hand, relatively little conceptual work has been done on the problem of optimizing the choice and frequency of monitoring tests. This is still an area where doctors must "fly by the seat of their pants."
   Consideration of whether or not a test is a suitable end point for a clinical trial is an informal blend of technical approaches to the classification problem and whatever testing practices happen to be in fashion in clinical practice at the time. This is not to say that many elegant and powerful types of study design and analysis strategies have not evolved from the partnership between physician-investigators and biostatisticians. In many cases, however, the statistical experts were forced to work with the end points the doctors prescribed.
   In the traditional medical formulation, "hard data" seem to come only from tests which involve high technology, which have become familiar to clinical policymakers in major medical centers. Perhaps it is not too cynical to observe that one standard for judging hard data is that their acquisition is eligible for third party reimbursement in fee-for-service health care systems. "Soft data," in contrast, are often obtained verbally, or by patient's diaries. They are cognitive rather than procedural, depending on people as filters. Their collection is not usually reimbursable.
   In the opinion of this writer, the distinction between hard and soft data as drawn above and exemplified by much of the clinical literature is inappropriate. It is based on a somewhat naive view of where useful information comes from and scientific generalizability. Clinical medical science has much to learn from the social sciences and other disciplines where scientific inference and testing theory have been more highly developed.
   For example, a nominally good test in traditional clinical research must be felt to have adequate sensitivity and specificity. Assessment of these parameters, however, means that some form of "gold standard" must be available, against which the test can be evaluated. For rapidly fatal conditions, patient survival is not a bad gold standard to use. But since chronic diseases, which kill slowly and have a major effect on quality of life, are now a central concern, the selection of adequate gold standards is usually problematic.

Accuracy

Precision

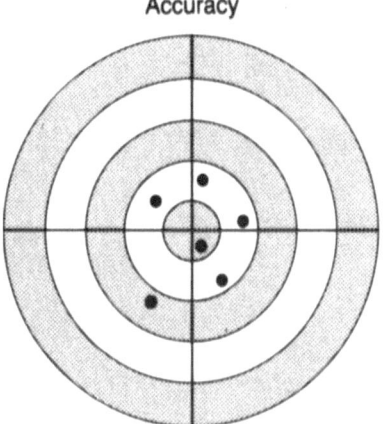
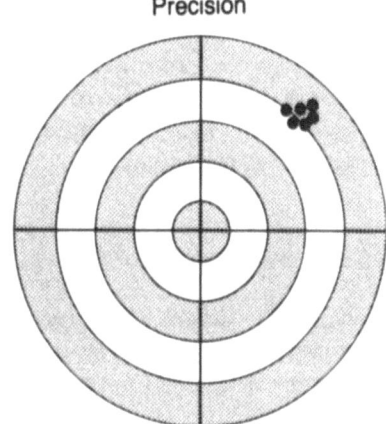

Fig. 7.3. The value of accuracy over precision

Clinical researchers tend to look to tests of basic biochemical of physiological function to bail them out of the gold standard problem. This may be a fine solution when the experimental line of inquiry has to do with understanding mechanisms of disease. They may not be appropriate as definitive end points in clinical trials intended to help show whether or not therapy is effective and should be promulgated. That is because correction of defects at the microlevel of functioning is not the object of health care – these processes are of interest to decision-makers because they predict (cause) changes in more global levels of functioning.

Without an adequate conceptual basis for dealing with imperfect gold standards, selection of clinical trial end points is likely to miss the target. A graphic presentation of the aphorism "It is better to be approximately right than precisely wrong" is shown in Fig. 7.3. For a decision-maker who must choose among alternatives, relevant, but noisy, data may be more useful than highly reliable information on an end point of only tangential interest.

Social scientists are familiar with the problem of measurement in the absence of a gold standard. They have evolved elegant techniques based on an exploration of the correlations between the measure of interest and measures thought to represent other related constructs. These tests of construct and discriminant validity can be used successfully to evaluate health status measures. [For a recent review and report, see Read JL, Quinn R, Hoefer MA [Measurement of overall health. J. Chronic Dis.] Further discussion of techniques borrowed from sociology and experimental psychology would not only help doctors think more clearly about socioeconomic measures but might also enlighten their approach to traditional medical end points.

Show Correlations Between New and Old Measures

Part of the construct validation approach is to show how newer measures stack up against more familiar end points. Acceptance will be enhanced by collecting and skillfully presenting data which show that good quality of life data move in the same direction as a variety of traditional medical tests. Qualitative comparisons can also help. A complex, difficult to learn, time-consuming quality of life measure which has nevertheless been shown to be valid can be justified on the same basis as a demanding medical test, so long as the distinction between testing for patient care and research purposes is kept in mind.

Perform and Report Model Studies

Because the methods are relatively unfamiliar and standards have not been well established, sponsors and investigators may occasionally be tempted to take short-cuts in the design and analysis of socioeconomic evaluations. This is likely to be counterproductive, since clinical decision-makers are reasonably sophisticated in their thinking about study design and statistical analysis. Publication of shoddy studies purporting to demonstrate quality of life benefits or cost-effectiveness can only delay the acceptance of these end points by clinical policymakers.

Methodological soundness does not mean that every socioeconomic evaluation must be based on a randomized, double-blind, placebo-controlled crossover design. Other research strategies may be called for, depending on the research question to be answered. Since these evaluations are often concerned with the effects of therapy in real-world applications, precision of experimental control may be deliberately sacrificed in order to obtain patient samples which are generalizable to the larger population.

Open-label designs may be appropriate if placebo and other patient factors (such as compliance) contribute to effectiveness. In the interests of maximizing credibility of such studies, however, the rationale for their design should be carefully explained. As with any evolving methodology, there is also a valid role for pilot studies, which, if reported, should carry the appropriate caveats.

New Roles for Clinical Policymakers

Part of the changing face of health care is the growing involvement of physicians in management roles. Assumption of responsibility for allocating resources will profoundly influence their interest in socioeconomic evaluations. When clinical policymaking becomes a formal responsibility, practitioners become motivated consumers of data on how well a particular therapy gives value for money.

Doctors do not need to run large health care organizations to become involved. Merely sitting as member of a Pharmacy and Therapeutics Committee or other group experiencing pressure from a hospital administration to contain costs will raise awareness. Professional medical input is essential for the deliberation of groups making decisions about what therapy will be available. An important bonus of their participation is that clinical policymakers become educated about the information needed for allocation of limited resources.

Conclusion

Socioeconomic evaluation in health care is still a young discipline. It has arisen in response to new but durable trends in the way society deals with the health care enterprise. For quite understandable reasons, many clinical policymakers have been slow to accept and contribute to the development of more sophisticated approaches to measuring quality of life and economic costs and savings.

Those interested in promoting further understanding of socioeconomic evaluation can accelerate the process by creatively merging their approaches with the traditional medical view of the world. It is worth the trouble to do so because decisions made by clinical policymakers still have the greatest impact in daily patient care.

8. Discussion of Part I

The presentations in Part I focused first of all on the actual economic milieu of drug therapy in the United States. The main topics of interest in this respect are
a) features of malpractice, including the way lawyers get paid in the United States;
b) competition and quality of care; and
c) cost containment.

The European contributors supplied additional views of the economic problems involved in drug therapy, namely
a) the pricing mechanisms in different countries, and
b) the quality of data on which the socioeconomic evaluation of drug therapy is based.

The peculiar feature of *malpractice* in the United States is a reflection of not having a welfare state. In the United States, people sue for the cost of medical care if the result is not what they had expected; therefore doctors stay away from new drugs if these prove risky. This tendency is supported by the way lawyers are paid in lawsuits for malpractice. The client (patient) bears no risk if the lawsuit is not successful; in case of success, the lawyer's share would be in proportion to the claim, one-third or so of the total amount. Lawyers, therefore, have an interest in pushing claims very high, a tendency that would not be considered ethical in other countries, especially those in Europe. Lawyers, however, may be very selective in choosing their cases, looking for winners. On this basis, they might end up having fewer cases. But, certainly, if one is being paid 30% of the gross award, this encourages higher claims, since the lawyer benefits as well as the patient (client).

In such a climate the providers are focusing their attention on quality assurance. The *competition* among the providers is extensive and ranges from more formal efforts to look at specific types of treatment, to documenting why some treatments do so well. Some institutions, e. g., hospitals, have formed committees of senior physicians to look at this question. Such committees are composed of the most senior, most influential physicians in the hospital. Other hospitals have taken a different approach by developing an elaborate computerized information system to monitor the costs of care (connected with the DRG payment mechanism) and the expectations regarding how well and how rapidly patients should recover. Such an information system keeps track of costs *and* of the quality of care.

Cost containment, which is still a big issue in the United States, is often promoted by health insurance companies. Working from articles published in leading medical journals, insurance companies define appropriate treatment for frequent diseases. In the future, they will pronounce explicit statements on what drugs should be used when, and in what quantities. The movement for decentralization of diagnostics and therapy will increase the interest in technology assessment. Since we are not likely to see a single authority vested with the right to make a uniform decision on cost-effective technologies, there will be a substantial interest in financing technology assessment by third-party payers, by employers, and by other groups.

The reality is that most of the troublesome technologies will not be so easily dispensed with. And it is precisely for this reason, and the inclination to have a diversity of opinions, that there will not be a centralized decision-making body saying yes, this technology is cost-effective, or no, this technology is not. However, there is some interest in the United States in trying to get some sort of a handle on the very wide variations that exist in practice style with regard to certain kinds of procedures and diagnoses. It is not clear how this is going to be implemented. A lot of attention has been given to the findings about the wide variations in surgical rates, for example, that exist for a certain diagnosis. This is particularly true for cases where the symptoms are unclear, the treatment given for the symptoms is unclear, or there is little agreed-upon information on the appropriate therapeutic strategy. While there has been much interest in documenting the variations, there is little agreement on what to do about them. Among other things, there is scant agreement on the appropriate level of treatment. The statement that usually follows the hand-wringing about the wide variations is that there is a need for clinical trials to determine, at least on a medical level, the more appropriate strategy, and then to figure out how to encourage it. In this example, one may see some movements to try to limit the extremes in the variations existing in the United States. There is a tendency not to head for the middle but to take what are the most diverse practice pattern associations and to cut off those extremes.

After coffee break, the discussion resumed first on pricing arguments. Obviously, the questions discussed centered around the *pricing of drugs* and the different prices for the same product in different countries. Regarding prices, one speaker pointed out that pharmaceutical innovation is a very risky business and that only one in eight products seems to come to any sort of fruition. But the question remains whether this kind of risk is any different than that of other industries, for example, computers or the steel industry. An industry representative tries to downgrade the risk of phar-maceutical R & D. In the past, the pharmaceutical industry has been a discovery-oriented industry ("to discover is always much more risky than to invent"), but it has slowly and gradually moved into the field of the innovation/invention process. Today, industry can look out for needs and decide to invent the necessary tool, almost to measure. It does not have to produce 10000 substances until one gets to the market. The risk lies in incomplete knowledge about the pathology and biological or pathophysiological mechanisms. Once they are known, modern industry can cope with it.

The difference in price in different countries does not reflect differences in R & D expenditure but rather the economic conditions within these countries. The real test of whether drugs are expensive in the country concerned is to compare them with competing modes of treatment. For example, in Italy hospital rates are inexpensive; thus, if one wants to compete against the hospital rate with a new drug, it needs to have a low price. However, in Germany where the hospital rates are expensive, a drug alternative will also be priced higher. The real test would be whether these ratios are approximately constant across European countries. Because hospital services are nontrade and drugs are traded, they should adjust to the relative prices across countries. Whether this is the case remains to be demonstrated.

Regarding the fundamental question of *properly measuring or evaluating,* one must set up clear-cut standards. In the socioeconomic field one cannot avoid dealing with what natural scientists would term "soft data." Socioeconomic audiences and researchers are very comfortable with such data. Of course, soft data need to be reproducible and valid, the instruments have to be reliable and validated, etc. The softness is more of a concern to the medical community and, in some cases, to the policy community, which are both more critical, of such data.

Referring to the area of cardiology, many cardiologists are obsessed by their machines and think just because they have a machine it must be objective. If the data are examined carefully, many machines are shown to be unreliable and the results not reproducible at all.

Following up, it seems much easier to term this part of the session a discussion about "good" data and "bad" data (rather than "soft" and "hard"). One *can* define what one means by data that are more or less persuasive than other data. Nevertheless, it is important to understand what people mean when they say hard and soft because these terms are *very* common in cardiology. Therefore, it is important for those of us who want to communicate with cardiologists to know what they view as hard and soft and, in fact, why.

One must admit that we are still lagging behind the psychometricians conceptually. One first of all has to think about identifying the variables in which one is interested. Second, one can have some a priori hypotheses regarding which indicator may be the basically good one, and the others might be in secondary roles.

Finally, we have to separate what we are measuring from how we are measuring. How we are measuring is at least as important as what we are measuring because whether the data one has are soft is really a function of how we measure. In fact, if we measure using an instrument that is well-conceived and well-documented, that came from well-controlled studies, and that has a high degree of reliability and validity, then we have taken the first step toward obtaining good data.

Part II

Measuring the Effectiveness of Drug Therapy –
The Economic Criteria

9. Cost-Effectiveness Study of Pharmaceuticals: Methodological Considerations

B. R. LUCE

Introduction

Pharmaceuticals have always been judged on the promise of effectiveness in treating, preventing, or ameliorating disease. Through the years, the standards against which drugs are measured have increased.

In 1906, the Food, Drug and Cosmetic Act established the new Food and Drug Administration (FDA) to regulate the marketing of drugs and foods for safety. It was passed in response to unsafe and falsely advertised products, primarily of the home remedy or "patent medicine" type. In 1938, an amendment requiring safety of drugs to be demonstrated through more rigorous and sophisticated testing was enacted, largely due to public catastrophes such as the Elixir of Sulfanilamide deaths [10]. In 1962, further drug amendments extended the FDA's regulatory authority to require that drugs be tested for efficacy as well as safety prior to marketing. Passage of this legislation was due, in large part, to the thalidomide tragedy [11]. As we approach the latter part of the 1980s, the drug industry is facing further demands – regarding cost-effectiveness – which sometimes include measuring the impact that drugs have on patients' overall quality of life.

This paper reviews briefly why and how these new demands are being made, and how they are being met within the American health care system with the realization that the same demands are being made, to differing degrees, throughout the Western world. This paper will cover the following broad topics:

1. The changing nature of health care financing and delivery in the United States and how these changes are affecting the market for pharmaceuticals
2. A brief overview of the technology assessment activities in the United States largely as a result of the above
3. Responses by the pharmaceutical industry in terms of testing their products for cost-effectiveness, including overall effects on patients' quality of life

The Changing Nature of Health Care Financing and Delivery in the United States

Medical care delivery and financing are undergoing profound changes which are deeply affecting the incentives and decisions of health care policy-makers, managers, providers, and consumers. Gone are the days, ushered in by first commercial and later government medical insurance, when the purchaser and user were largely insulated from the financial consequences of his or her decisions. Medicare is paying hospital care prospectively on a per case basis, which encourages hospital administrators to minimize resources used per patient. Medicare is also encouraging capitation arrangements for the elderly, which stimulates resource constraints for all acute care. State Medicaid agencies are restricting drug formularies, experimenting with managed care systems, and selectively contracting with providers. Private industry is self-insuring and throughout the private sector we are seeing discounting practices, e.g., in the form of preferred provider organizations, capitation arrangements, and both vertical and horizontal integration of financing and various levels of delivery arrangements.

This new competitive environment is causing the purchasers of health care to question more carefully the value of all resources purchased, including pharmaceuticals. Thus, the pharmaceutical industry has responded by sponsoring cost-effective studies of its products to meet these new market demands. Purchasers want to know not only if a product has value (i.e., whether it is efficacious), but also whether it is cost-effective.

The Changing Nature of Technology Assessment Activities in the United States

The expansive environment of the 1960s which was designed to increase access to health care and which led to the passage of Medicare and Medicaid, gave way to the decade of regulation of the 1970s to contain spiraling costs. The medical products industry, and the health care system itself, was in a mode of expansion, innovation, and experimentation. But government became concerned with the cost of health care and sponsored a number of regulatory activities designed in part to contain the technological expansion. Health planning laws were enacted, hospital utilization review organizations were established, and limits were placed on reimbursement. In addition, research funds were invested to improve assessment techniques. For instance, the National Center for Health Services Research sponsored a number of studies to develop and improve the state-of-the-art of cost-effectiveness analysis and the measurement of both health status and quality of life.

In the early 1970s, Congress established the Office of Technology Assessment (OTA), whose Health Program sponsored several studies examining the state-of-the-art and policy usefulness of various assessment techniques, including cost-effectiveness analysis. One of the studies examined the cost-effectiveness of pneumococcal vaccine for the Medicare program [13]. This study is described later in this paper.

In 1978, Congress also established the National Center for Health Care Technology (NCHCT), whose primary missions were to assess existing, new, and emerging technologies and to advise the Health Care Financing Administration in its coverage decisions. It was widely believed that the new Technology Center would help slow the seemingly insatiable appetite of the United States health care system in consuming technological innovations.

Since the government was clearly looking to technology assessment as a means to constrain health care costs, few of these efforts were encouraged by either the health professions or the medical products industry. In fact, due to a loss of private political support, NCHCT was defunded in 1981. But the new financial incentives of the 1980s, such as Medicare's DRG-based Prospective Payment Systems, changed much of that. Since the incentives themselves threatened to control technology expansion in the health care arena, industry became more supportive of technology assessment efforts. Instead of assessment as a means to contain the development, adoption, and use of technologies, industry began to view assessment as a legitimate way to demonstrate the value of their innovations. Thus, whereas the *demise* of NCHCT was supported by industry in 1981, the *creation* of the Medicare Prospective Payment Commission was supported by industry in 1983, as was the Council of Health Technology at the Institute of Medicine and the National Advisory Council to the Office of Health Technology Assessment in the Public Health Service. More important for the purposes of paper, however, industry began broadening its own technology assessment of its products in the form of cost-effectiveness analysis, including the measurement of the effects which their products may have on the quality of patients' lives.

The Pharmaceutical Industry Response

By 1980, the Pharmaceutical Manufacturers Association (PMA) embarked on an initiative to assist its member companies in understanding the methodology and value of cost-effectiveness analysis of pharmaceuticals. A number of papers and studies were prepared for PMA by both internal and external research analysts, culminating in the publication of PMA's "Blue Series." Among the publications were general methodological guidelines for conducting cost-effectiveness studies of pharmaceuticals and three papers analyzing the cost-effectiveness of beta blockers for angina, glaucoma, and the prevention of the recurrence of myocardial infarction [1–3, 6].

The purpose of the PMA's initiative in cost-effectiveness was twofold. First, to demonstrate that pharmaceuticals are often a cost-effective substitute for medical and surgical therapy, and second, to help its member firms learn the technique in the belief that these studies would be increasingly required by the marketplace.

Efforts by individual firms date back to at least 1977, when Smith, Kline, and French initiated a whole series of cost-effectiveness studies to support the marketing effort of its (then) new anti-ulcer drug, cimetidine (Tagamet). Mainly, these studies documented that Tagamet saved insurers money by decreasing the need for ulcer surgery, and they complemented the many clinical studies demonstrating safety and efficacy [9].

Today, firms throughout the pharmaceutical industry are sponsoring such studies. Particular attention is being paid to making such studies an informative, scientifically valid adjunct to a firm's product marketing strategy. The following issues are being paid particular attention:

The Framework of the Analysis

Perspective

Studies are being tailored to be relevant to the particular audience of interest. The question that is asked is "for whom is the drug supposed to be cost-effective: the hospital, insurer, patient, or society"? Addressing the relevant perspective helps determine what costs and what benefits are to be counted, as well as how to count them and how to design the study for maximum usefulness. To illustrate the importance of perspective, consider that a cost to an insurer is a revenue to a hospital. Thus, an innovation which decreases hospitalization is one which benefits an insurer. In the cimetidine case, total Medicaid costs were calculated for similar ulcer patients before and after the drug was introduced (Geweki).

Comparison

Although the Food and Drug Administration mainly requires that a new drug be compared to a placebo, the marketplace generally wants a comparison to the relevant alternative, whether it be another drug, surgery, medical treatment, or nothing at all. Cimetidine, for instance, was compared to surgery. Today, there are many examples of "head-to-head" competition between drugs [5, 8, 12] or between other therapies [7, 14, 15].

Study Independence

Numbered are the days when cost-effectiveness claims can be made by company sales persons without independent scientific studies to back the assertions. Hospitals, Health Maintenance Organizations, and Medicaid agencies are becoming increasingly more sophisticated about study design and analysis. Pharmaceutical firms are increasingly turning to academicians [5] and contract research firms [7, 8, 12, 14, 15] and thus beginning to treat cost-effectiveness studies as they do their clinical efficacy research.

Research Design Issues

There are a number of study designs which are appropriate depending upon the research questions and the resources available (Fig. 9.1).

Probably the most common type of industry-sponsored study found in the health economic literature today is a simple *cost–cost* comparison. That is, comparing the cost of treating some medical condition using two types of therapy. This design normally assumes that both therapies are equally effective. One example of a cost –cost study is that by Eisenberg et al. [8], in which a new cephalosporin (Monocid) was compared to standard intravenous antibiotic therapy. Monocid's main advantage was claimed to be efficiency: it could be administered once per day instead of three or four times per day. The authors assumed there was no therapeutic advantage or disadvantage of the two drugs if taken as prescribed. Thus, the study consisted of carefully

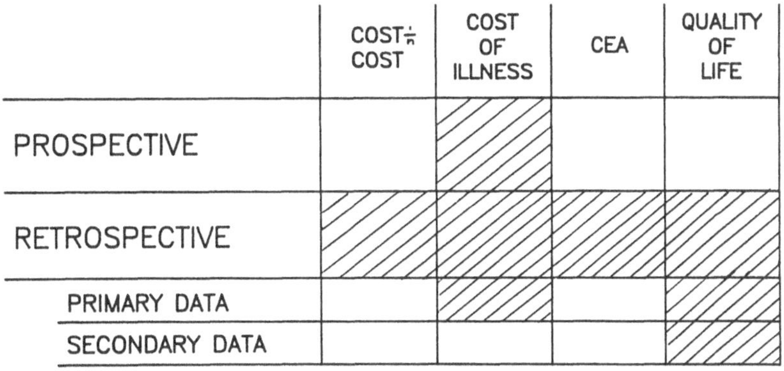

	COST÷COST	COST OF ILLNESS	CEA	QUALITY OF LIFE
PROSPECTIVE		////		
RETROSPECTIVE	////	////	////	////
PRIMARY DATA		////		////
SECONDARY DATA				////

Fig. 9.1. Types of study. *CEA:* cost-effectiveness analysis

measuring production costs associated with the purchase of the drugs, equipment, and supplies and the personnel time it takes to set up intravenous infusions and monitor the drug administration. The authors reported savings of over $ 5.00 per day per patient. Another cost–cost study, performed [12] compared the use of nifedipine, an orally administered calcium channel blocking drug, to standard intravenous therapy (mainly nitroprusside) for the treatment of hypertensive emergencies. Clinical tests were indicating that nifedipine was as therapeutically efficient as nitroprusside, evidently without requiring intensive nursing monitoring due to the potency of nitroprusside. The study consisted of retrospectively matching patients with clinical characteristics from different hospitals and monitoring the level of hospitalization that was incurred by the respective groups. Substantial savings were calculated for the nifedipine group because these patients were less apt to be admitted to the intensive care unit and less apt to be assigned to a full-time nurse for monitoring purposes.

Cost–cost studies can be either prospective or retrospective. As an alternative, net cost can be compared with the net health effects gained, which is the classic cost-effectiveness analysis. If all effects are valued economically, the study is termed a cost-benefit analysis. Cost-benefit analyses are becoming less common partly because of the social disagreement about valuing life and partly because it is difficult to argue convincingly that some relevant party reaps that value. The pneumococcal vaccine study, mentioned earlier, is a good example of a cost-effectiveness analysis, where the vaccine was compared to no vaccine [13]. In this retrospective, simulation study, the total cost to the nation of pneumococcal pneumonia was determined and then compared to the total costs of vaccination and treatment of the disease. The net cost difference was then juxtaposed to the expected net lives and morbidity savings due to the vaccination program. The results were expressed in net costs per quality adjusted life years saved. Results indicated that it cost between $ 77000 (for childred 2–4 years of age) and $ 1000 (for adults over 65) per year of healthy life. The OTA-sponsored study resulted in a change in legislation, with pneumococcal vaccine now being the first (and only) purely preventive benefit in the Medicare program. A similarly designed retrospective study [14] calculated the expected cost-effectiveness of chewing nicotine gum as an adjunct to physician's counseling to stop smoking. Using published clinical data regarding smoking cessation rates, and similar available information concerning the relationship between smoking, health, and the cost of disease, the authors calculated that the use of nicotine gum results in a cost per year of life saved from $ 4113 to $ 9473.

Finally, some studies are beginning to appear in the literature that compare only effects on quality of life, which we term here "effect–effect" studies. Although not *cost*-effectiveness studies, in that no costs are involved, these studies can be considered within the general boundaries in this paper in that they are being employed due to market pressures and they extend the scientific assessment of pharmaceuticals beyond the traditional one of safety and efficacy. Also, they can be integrated into cost-effectiveness studies.

Measurement of quality of life, sometimes referred to as health status, is still being developed but is coming to mean the use of one of a number of broadly focused questionnaires associated with general "domains" of health such as activities of daily living, psychosocial activities, or general feelings of well-being (see Fig. 9.1). Other, narrower domains may center on feelings of anxiety, depression, or sexual dysfunction. What most of these techniques have in common is that they tend to be broader-based than typical symptom and side-effects measurement, the questionnaire can usually be scored in aggregate, and they can be used for multiple, if not most, diseases and interventions.

There are a number of examples of quality of life measurement studies in the literature, most of which are prospective since questionnaires are needed. Such information is normally not routinely recorded in medical records or claims files. The study which is already becoming a classic in the literature is one by Croog et al. [5] that measures the effects of differing hypertensive therapy. Using measures such as general well-being, sexual dysfunction, work performance, life satisfaction, and social participation, the authors were able to demonstrate, in a randomized controlled clinical trial, significantly higher levels of quality of life for patients prescribed one antihypertensive drug (captopril) than for those prescribed two others (propranolol and methyldopa).

Another study by Bombardier et al. [4], compared auranofin (oral gold) and placebo in a randomized, double-blind study for the treatment of rheumatoid arthritis. Along with traditional efficacy measures such as number of tender joints, nontraditional "quality of life" measures were assessed, including functional performance, pain, global impression, and utility (worth or value). The study found auranofin to be superior to placebo in the clinical, functional, and global composite scores.

Table 9.1. Assessing quality of life

Functional capacity	Perceptions
■ Daily routine	■ Health status
■ Social	■ Well-being
■ Intellectual	
■ Emotional	
■ Economic	Symptoms

Prospective vs. Retrospective Studies

Depending upon the availability of data, time, and finances, studies can be either prospective or retrospective. Prospective studies are usually more expensive and take longer, but they provide the opportunity to use rigorous, randomized designs and thus can produce very high quality analyses. They also afford the opportunity to measure effects on quality of life (since patient questionnaires are needed), which is usually impossible using a retrospective design. However, if cost-effectiveness is included within a planned phase III or IV clinical trial, cost may not be high and time is not necessarily an issue, since the results will be known at the same time the clinical trial is completed. At least in one recent case, FDA required a quality of life component in a phase III trial.

Retrospective studies are more commonly used for a number of reasons. They tend to be faster and less expensive, as noted above. Another reason may have to do with the newness of this kind of research within the pharmaceutical industry. The marketing departments of firms are becoming aware of the usefulness of these studies ahead of the clinical research departments. Therefore, today these studies tend not to be considered until after the drug has reached the market.

Retrospective studies can use both primary and/or secondary data. Primary data could be medical records or, more likely, medical claims data files. Secondary data normally is a synthesis of what is already in the literature.

In practice, many studies use both, augmented with expert opinion, for instance from physician panels, to fill in data gaps.

Presentation of Findings

The main vehicles used to present the findings of these industry-sponsored cost-effectiveness studies is coming to be the peer-reviewed professional literature. Dow's nicorette study was published in *JAMA* [14]; Squibb's quality of life study of captopril was published in the New England Journal of Medicine [5]. Smith, Kline, and French's studies of auranofin [4] and monocid [8] were published in *The Journal of American Medicine* and *Reviews of Infectious Diseases*, respectively. Such a trend is noteworthy because, in the past, cost-effectiveness claims have often resulted from unsophisticated studies performed either in-house or by consultants and the dissemination vehicle has been limited to company brochures and other proprietary sales media.

Articles in the professional literature carefully lay out the assumptions, the methodology, the strengths and weaknesses of the study, and the generalization of the results. Sensitivity analysis is commonly done on important assumptions and variables that have uncertain values.

Summary

With the more competitive environment in the United States, the industry seems to be responding appropriately by devoting more resources to cost-effectiveness studies of their products. Studies are so designed as to be targeted to the specific perspective of the consumer to be reached. The cost-effectiveness of various pharmaceuticals is being compared, and independent academics and research firms are being utilized.

Design of studies take on a number of forms, including comparisons of cost, effects, or both, and they may be prospective or retrospective analyses. In addition, studies are attempting to measure effects of pharmaceuticals and other interventions on the general quality of patients' lives.

It is expected that as firms become more aware of the usefulness of such research, cost-effectiveness, including the measurement of quality of life, will become common during phase III clinical trials.

References

1. Bentkover JD (1984) Beta blocker, reduction of mortality and reinfarction rate in survivors of myocardial infarctions, Pharmaceutical Manufacturers Association, Washington, D.C.
2. Bentkover JD (1984) The use of a beta blocker in the treatment of glaucoma: a cost-benefit study. Pharmaceutical Manufacturers Association, Washington, D.C.
3. Bentkover JD (1984) The use of beta blockers in the treatment of angina: a cost-benefit study. Pharmaceutical Manufacturers Association, Washington, D.C.
4. Bombardier C, Ware J. Russel IJ, Larson M, Chalmers A, Read JL (1986) Auranofin therapy and quality of life in patients with rheumatoid arthritis: results of a clinical trial. J Am Med 81:565–578
5. Croog SH, Levine S, Testa MA et al. (1986) The effects of antihypertensive therapy on the quality of life. N Engl J Med 314:1657–1664
6. Dao TD (1984) Cost-benefit and cost-effectiveness of pharmaceutical intervention. Pharmaceutical Manufacturers Association, Washington, D.C.
7. Eisenberg JM, Kitz DS (1986) Savings from outpatient antibiotic therapy for osteomyelitis: economic analysis of a therapeutic strategy. JAMA 255:1584–1588
8. Eisenberg JM, Koffer H, Finkler SA (1984) Economic analysis of a new drug: potential savings in hospital operating costs from the use of a once-daily regimen of a parenteral cephalosporin. Rev Infect Dis 6 [Suppl 4]:909–923
9. Geweki J, Weisbrod BA (1984) How does technological change affect health care expenditures? The case of a new drug. Evaluation Res 8:75–92
10. Lambert EC (1978) Modern medical mistakes, Indiana University Press, Bloomington
11. Lawless EW (1977) Technology and social shock, N.J. Rutgers University Press, New Brunswick
12. Luce BR, Ellrodt AG, Cameron JM, Reidinger M (1985) Managing acute hypertension: cost considerations. Am J Emerg Med 3:6 [Suppl]
13. Office of technology Assessment, U.S. Congress (1979) A review of selected vaccine and immunization policies. Government Printing Office, Washington, D.C.
14. Oster G, Huse DM, Delea TE, Colditz GA (1986) Cost-effectiveness of nicotine gum as an adjunct to physician's advice against cigarette smoking. JAMA 256:1315–1318
15. Oster G, Tuden RL, Colditz GA (1987) A cost-effectiveness analysis of prophylaxis against deep-vein thrombosis in major orthopedic surgery. JAMA 257:203–208

10. The Impact of Angina Pectoris on the Patient and the Influence of Treatment

S. H. TAYLOR

Recent years have witnessed a tremendous expansion of knowledge of the pathophysiological mechanisms underlying the syndrome of angina pectoris. Following Heberden's classic description of the clinical syndrome in 1772, nearly two centuries were to elapse before the pathological features underlying the precipitation of ischaemic cardiac pain were clarified. The development of modern methods of study in the experimental animal and safe methods of investigation in man were major contributors in this regard. Such advances were also stimulated by the increasing frequency of presentation of the condition during the past half century. It is perhaps difficult to realise that such a popular and fine diagnostician as Sir William Osler (1910) found angina to be amongst the least frequent of the cardiovascular conditions he was called upon to treat at the turn of the century. A third major factor contributing to the intense clinical interest in this condition has been the development of drugs and surgical methods of relieving symptoms in the angina patient.

It is unfortunate, but perhaps understandable, that the of concentration on the underlying pathophysiology and its clinical sequelae has tended to distract from the consequences of angina pectoris on the patient. In the search for a remedy for the ominous symptoms, physicians in the past have perhaps too readily ignored the twin impact of such sinister occurrences and such potent treatments on the quality of life the patient afflicted by angina pectoris. Despite the difficulties of measuring such effects, this new and important aspect is now receiving increasing attention [14].

Pathophysiology of Angina Pectoris and Its Clinical Consequences

The pathophysiological background of angina pectoris in relation to treatment has been reviewed previously [12]. The syndrome of angina pectoris results from inequality of myocardial oxygen demand and coronary blood supply. Under normal circumstances there is a direct and linear relationship between myocardial oxygen consumption and coronary blood flow. Imbalance may result either from restriction of the coronary blood flow or from an unbalanced increase in myocardial oxygen consumption. Undoubtedly a combination of both factors is by far the most frequent cause of ischaemic cardiac pain, although in the individual patient it is difficult to apportion the

proportionate magnitude of the pathophysiological mechanisms involved in the production of symptoms.

Restriction of coronary blood supply may result from organic lesions partially
occluding the lumen of the large coronary vessels. Single or multiple atheromatous
depositions are almost invariably the source of such organic obstruction. However, it
is now realised that some of the waxing and waning of the disease syndrome may also
be due to other factors. Functional narrowing of the coronary arteries may occur in
the vicinity of an atheromatous obstruction. Transient superimposition of blood
clotting elements on the atheromatous plaque may also be a cause of anginal aggrava-
tion.

The oxygen consumption of the myocardium is dependent upon many physical and
metabolic factors but in the patient with angina pectoris the major factor causing
myocardial ischaemia is undoubtedly sympathetic stimulation of the heart. This
results in incrase in heart rate, systemic blood pressure and myocardial contractility,
all of which create an additional oxygen demand sufficient to precipitate anginal pain.

It is thus possible in broad pathophysiological terms to describe angina as caused by
a predominant deficiency in coronary blood supply and, consequently, by appropriate
myocardial oxygen demands. This is undoubtedly a useful concept with definite
therapeutic implications. Unfortunately the difficulty of deciding which factors pre-
dominate in the production of anginal pain in the individual patient limits its extrapo-
lation to an exact pharmacological remedy. The majority of patients with angina still
obtain relief from their oppressive symptoms by trial and error treatments often
involving different drug combinations.

The development of anginal symptoms carries a sinister prognosis for the patient.
The mortality risk is directly related to the extent of involvement of the major
coronary vessels (Fig. 10.1.). Moreover coronary artery disease is a progressive
syndrome and in the patient presenting with ischaemic cardiac pain the rate of
progression, even of initially undamaged vessels, greatly exceeds that in the normal
population. Over a 2- to 3-year period, 5% of coronary arteries with no detectable
narrowings at the first study will be more than 50% narrowed at the time of the second
angiogram. Approximately 10% of arteries with a recognisable narrowing at the first
arteriogram will have further narrowing of this lesion by at least 40% 2–3 years later.
Ominously, new total occlusions can be expected in approximately 13% over the same
period [4]. Thus whilst the practitioner may be vaguely aware of the serious nature of
his patients's illness, all too frequently his therapeutic approach does not reflect the
urgency it warrants. Moreover, due to the publicity that angina and its treatment
frequently receives, patients are becoming increasingly aware of their problematic
prognosis.

The Impact of the Disease Syndrome on the Patient

How does the advent of symptoms of ischaemic cardiac pain affect the patient's
"quality of life"? Undoubtedly, recurrent attacks of anginal pain restrict the patient's
ability to live and enjoy a normal life. Unfortunately, the complexities involved in
attempting to measure such an abstraction are only now receiving the attention they
warrant [14]. On first acquaintance the meaning of "quality of life" seems quite clear;

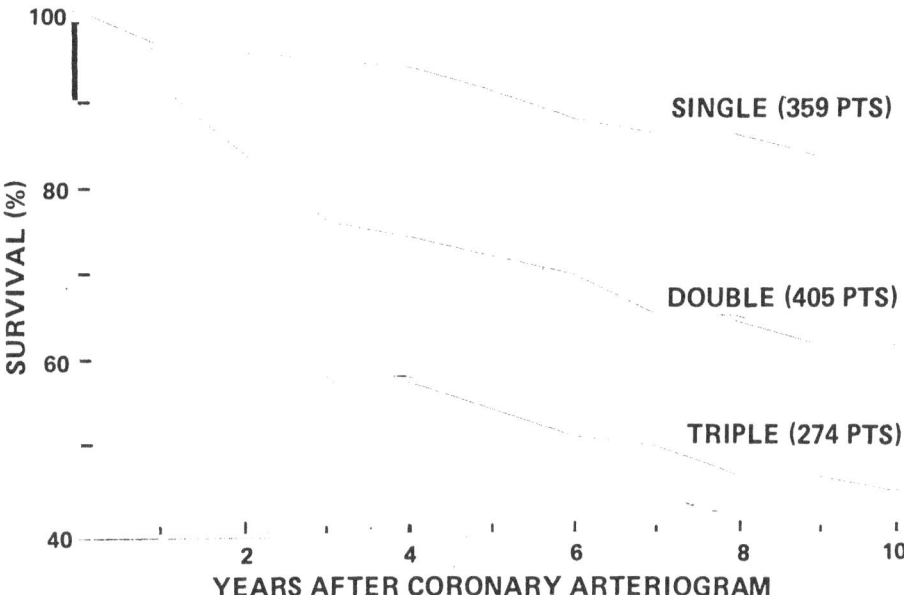

Fig. 10.1. Patient survival related to number of coronary arteries narrowed > 50% (single, double or triple vessel disease) at angiography in 1038 patients combined from four studies in different centres. (Griffith and Achuff [4])

closer scrutiny, however, reveals how difficult it is to define it with convincing certainty. Quality of life is subjective, idiosyncratic and variable. The complex character and interaction between the physical, psychological and emotional factors that make life pleasurable for a given individual are exceedingly difficult to describe in acceptable qualitative and quantitative terms. For instance, quality of life cannot simply and directly be translated in terms of "happiness" [3]. This is an emotion that is affected by many factors other than physical health. Quality of life may also be described from a number of viewpoints. It may be described directly by the patient, it may be discerned by his close relatives, it may be perceived by friends and acquaintances or it may stem from his physician's overview. Which standpoint is the most appropriate is a matter of debate: probably all are relevant from different aspects of the patient's life.

Determination of the impact of angina pectoris on the patient can only be achieved with certainty by apportioning those aspects which can be objectively measured. In the patient with such serious symptoms as cardiac pain, his physical disability undoubtedly plays the dominant role in his life quality. The impact of his symptoms on his quality of life is likely to be substantial from two aspects. Restriction of physical activity and therapeutic preclusion from those exciting events that colour life can be expected to detract substantially from his quality of life. Fortunately it is easy to measure these relatively well-defined restrictions by objective means such as exercise testing. But can the total impact of the syndrome be determined from a wider viewpoint? How relevant are more subjective measures of quality of life in the patient with angina pectoris?

The factors that go to make up the quality of life in patients suffering from such sinister symptoms as recurrent anginal pain are obviously quite different from those factors that encourage enjoyment of life in the normal subject or one with less serious and less incapacitating symptoms. Thus assessment methods concerned with measuring the general quality of life of a patient are open to serious error when specifically applied to the patient with angina pectoris. Many instruments have been developed in an attempt to provide a comprehensive measure of the quality of life in patients afflicted by disease [11]. However, by their very nature their breadth reduces their specificity and validity when applied to such a distinct and sinister symptom complex as angina pectoris. At present it is reasonable to conclude that these general instruments, however useful they may be in non-sinister illnesses, probably have little general application in the measurement of this complex abstraction in patients with such a specific and serious symptom as angina pectoris. An instrument that is relevant, is valid, with a high discriminative power and has a high degree of repeatability is yet to be developed. Such an instrument must also be of sufficient sensitivity to reflect with fidelity changes in the natural history of the disease and to detect changes specifically related to the treatment administered. Such measurements based on subjective factors must be shown to be correlated with other more objective measures of the disease process. Moreover, weighting of the many components that contribute to the overall quality of life of the patient must be carefully determined.

A further consideration that can be expected to cause deterioration in the quality of life of many patients with angina pectoris is the anxiety–depressive reaction that many suffer due to knowledge of the nature of their illness and their shortened life span. Many go in constant peril of myocardial infarction and sudden death, threats well publicised by the media. Patients with such incapacitating and worrying symptoms as angina pectoris must also be greatly influenced by the availability and quality of the medical care available.

Thus the impact of angina pectoris on the patient has many facets. Undoubtedly the physical symptom of chest pain is by far the most important factor in inhibiting enjoyment of life, and is rightly the major target of therapy. However, many other factors contribute to the emotional and intellectual discomfort and aggravate the primary disability. Awareness of the impact of angina pectoris on the patient as a whole is critical for successful treatment.

Impact of Anti-anginal Treatment on the Patient

It is only in relatively recent times that the overall inherence of treatment has been taken into account in the patient with angina pectoris. Rightly, the overriding concern of the patient and his physician is the abolition of his symptoms. Whilst this may not substantially prolong his life, there is little doubt that it will dramatically improve the quality of that remaining to him. Thus the prime aim of treatment of the patient with angina pectoris is relief of his symptoms, hopefully in parallel with a reduction in objective evidence of myocardial ischaemia. But secondary considerations are also likely to have an appreciable impact upon the efficacy of treatment once such primary aims have been achieved.

The threat of side-effects looms large when using potent antianginal drugs. Quite different to the case obtaining in hypertension, treatment is mandatory in angina pectoris irrespective of the adverse influence on the patient's quality of life (always assuming that the drug affords a major reduction in anginal attacks). Naturally, intense effort has been devoted to minimising these adverse influences by better methods of drug delivery. But many factors other than direct drug therapy can be expected to exert an influence on the patient's quality of life. Expert and reassuring advice, particularly directed to relieving the patient's natural anxiety, and sensible advice regarding secondary preventive measures can be expected to give a distinct uplift to the morale of the patient. Improvement of the psychological well-being of patients has been reported in a number of clinical trials independent of the drugs administered [8]. The social and psychological support given by medical cover alone has been confirmed by the reductions in distress scores observed during attendance at medical clinics even before placebo was given [6]. Such influences further complicate an already confusing array of factors influencing the quality of life in patients with angina pectoris. Until the complex character of the numerous components that constitute quality of life in the angina patient are clear, substantial difficulties will remain in interpreting the secondary consequences of drug-induced success in relieving their primary symptoms.

Measurements of the influence of treatment on the quality of life of the patient with angina pectoris are sparse. Using a crude scoring technique it has been claimed that quality of life is improved in patients undergoing coronary artery bypass surgery for angina pectoris [2]. Many drugs, including nitrates, beta-blockers and calcium antagonists, have been shown to be highly effective in abolishing the symptoms of angina pectoris. Although there are as yet no reports of a major controlled study devoted solely to measuring the influence of antianginal drugs on life quality in patients with angina pectoris, large-scale open trials in general practice have indicated that transdermal glyceryl trinitrate patches improve crude measures of the quality of life [1, 7].

The effectiveness of glyceryl trinitrate in relieving anginal pain is beyond doubt. In view of the efficacy of the transdermal route of delivery and particularly by controlled release patch [9, 13], it is clear that such delivery systems have considerable potential not only in ameliorating anginal pain but also in reducing side-effects.

Nevertheless, it is now obvious that better controlled studies specifically designed to explore the impact of all anti-anginal treatments on the life quality of patients with angina pectoris are required. But these must await the development of a specific, sensitive, discriminative and valid instrument.

Conclusions

Angina pectoris has a major distressing impact on all aspects of the patient's life-style. Restriction of physical activity is undoubtedly the dominating feature, but other secondary factors may substantially aggravate the primary symptom-disorder. As yet there is no infallible method of measuring "quality of life" in the patient with angina pectoris.

Treatment of the patient with ischaemic cardiac pain is primarily directed to relief of organic symptoms. Many effective treatments are available. However, their impact on the quality of life of the patient, over and above that related to the relief of symptoms, is at present unknown and awaits clarification.

References

1. Bridgman KM, Carr M, Tattersall AB (1984) Post-marketing surveillance of the Transderm-Nitro patch in general practice. J Int Med Res 12:40–45
2. CASS Principle Investigators and Their Associates. Coronary Artery Surgery Study (CASS) (1983) A randomised trial of coronary artery bypass surgery. Quality of life in patients randomly assigned to treatment groups. Circulation 68:951–960
3. George LK (1979) The happiness syndrome: methodological and substantive issues in the study of social-psychological well-being in adulthood. Gerontologist 19:210–216
4. Griffith LSC, Achuff SC (1977) Coronary arteriography and left ventriculography. In: Julian DG (ed) Angina pectoris. Churchill Livingstone, Edinburgh, p 164
5. Heberden W (1772) Some account of a disorder of the breast. Med Trans R Coll Physicians Lond 2:59–63
6. Kellner R, Sheffield B (1971) The relief of distress following attendance at a clinic. Br J Psych 118:195–198
7. Letzel H, Johnson LC (1984) Therapy of angina pectoris with Nitroderm TTS. Results of a multicentre field study with 37,596 patients. Med Welt 35:326–334
8. Mann A (1984) Hypertension: psychological aspects and diagnostic impact in a clinical trial. Psychol Med (Monograph Supplement 5) Cambridge University Press, Cambridge 1–35
9. Muiesan G, Agabiti-Rosei E, Muiesan L et al. (1986) A multicentre trial of transdermal nitroglycerin in exercise-induced angina: individual response after repeated administration. Am Heart J 112:233–238
10. Osler Sir W (1910) The Lumleian lectures on angina pectoris. Lancet I:839–844
11. Storstein L How should changes in life-style be measured in cardiovascular disease. Am Heart J. 114, (1):210–212
12. Taylor SH (1982) Pathophysiological approach to the treatment of angina pectoris. In: Coltart J, Jewitt D (ed) Recent developments in cardiovascular drugs. Churchill Livingstone, Edinburgh pp 135–163
13. Taylor SH (1986) The role of transdermal nitroglycerin in the treatment of coronary heart disease. Am Heart J 112:197–207
14. Taylor SH Drug therapy and quality of life in angina pectoris. Am Heart J. 114, (1):234–240

11. The Socioeconomic Study Program on Nitroderm TTS

P. LAUPER

A new Drug Therapy Must Demonstrate Not Only Its Efficacy and Safety But Also Its Cost-Effectiveness

Costs of health care are rising all over the world, and the rate of increase is often greater than that of the gross domestic product and of the general cost of living (Fig. 11.1). This development is commonly referred to as a "cost explosion." It is manifested by a steadily growing burden on national economic systems, on the taxpayer, and on the patient. Today, instead of focusing on expansion of services and optimum health care those concerned speak primarily of curbing costs and of cost-effectiveness.

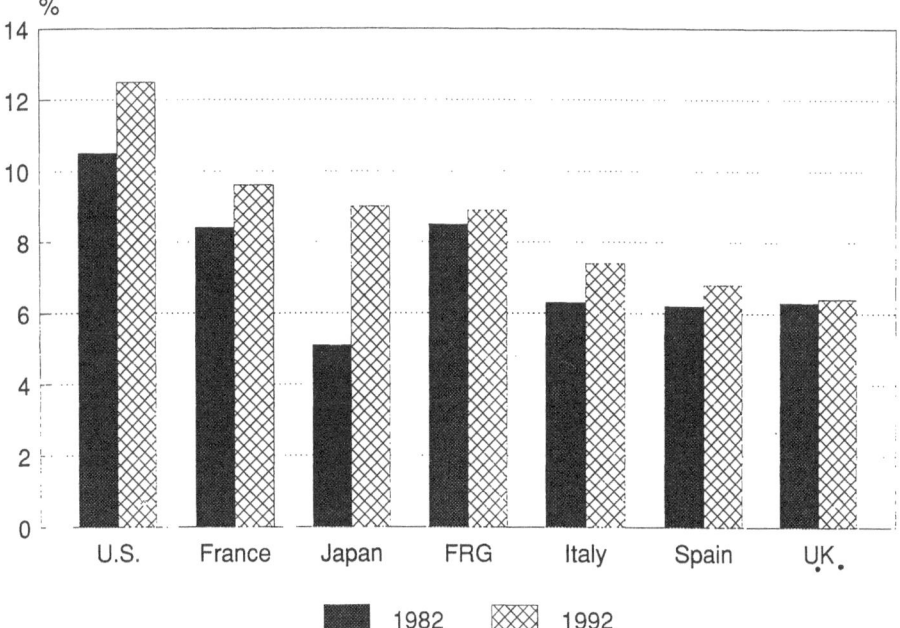

Fig. 11.1. Health care expenditures and economic growth

This particularly applies to the field of drug therapy: In this area not everything that is medically and technically possible for the elimination and reduction of illness is necessarily capable of being financed.

Advances in pharmacological medicine are likewise not immune to the pressure to conserve resources and increase cost-effectiveness. It is no longer enough to demonstrate the mode of action, efficacy, and safety of a new pharmaceutical product. One must show that the benefits obtainable from its use will bear a positive relationship to the costs it generates. In other words, a new preparation must demonstrate its cost-effectiveness.

Does Nitroderm TTS Fulfill the Cost-Effectiveness Requirements?

Ciba-Geigy introduced the first nitroglycerin transdermal therapeutic system in 1982, placing at the disposal of doctors, for the first time, a preparation that enables nitroglycerin to be administered through the skin in a controlled dosage for the long-term prophylaxis of angina pectoris attacks. The preparation immediately met with an enthusiastic reception and is now available in more than 50 countries and used by about a million patients every day.

The costs per pack and per dose of Nitroderm TTS are higher than those of traditional nitrate preparations in tablet, capsule, or ointment form. This price difference is a result of the development and production costs and the sophisticated pharmacological formulation. However, this difference can be justified only if the preparation gives greater benefits. Ciba-Geigy therefore set about conducting empirical studies at an early stage to provide evidence substantiating the product's value for money. Today, Nitroderm TTS is from the economic viewpoint one of the most thoroughly analyzed innovations in drug therapy. The studies confirm that, despite the per dose price difference, Nitroderm TTS fulfills the justified cost-effectiveness demands.

Rationale for the Cost-Effectiveness of Nitroderm TTS

The cost-effectiveness of pharmacological therapies cannot be measured by comparing their costs in isolation. On the contrary, a meaningful assessment can only be made if, in addition to these costs, the economically relevant effects of using the preparation are taken into consideration as comprehensively as possible. The spectrum of these effects ranges from impact on the use of concomitant medication, medicotechnical services, and doctors' services, through effects on the number of ambulatory or hospitalized days of illness and absence from work to qualitative influences on the mental, physical, and social well-being of the patient.

Several premises underline the cost-effectiveness analysis of Nitroderm TTS. The following hypotheses have been advanced (Fig. 11.2):
– By comparison with conventional long-acting oral nitrates, Nitroderm TTS has the advantages of higher patient compliance, an enhanced placebo effect, the avoidance of first-pass metabolism in the liver, and release into the body of a controlled quantity of nitroglycerin.

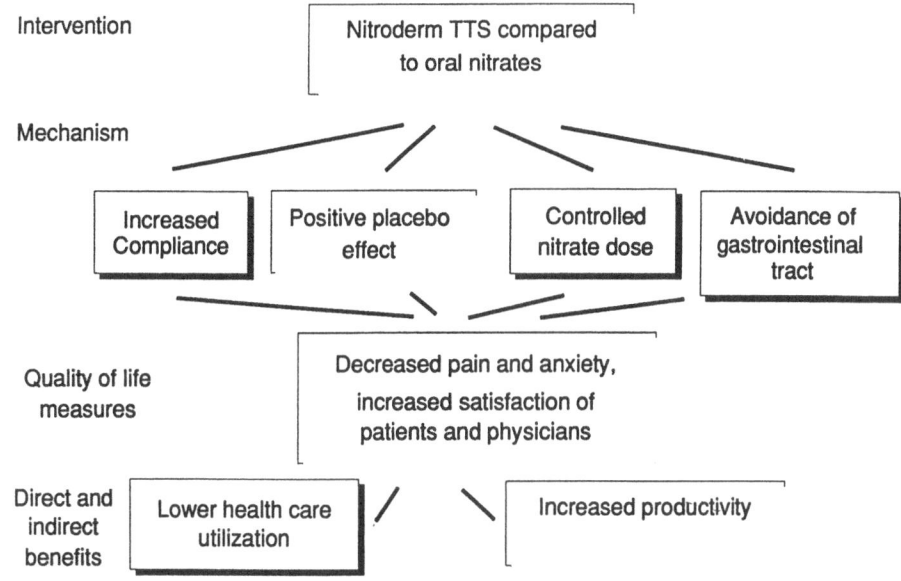

Fig. 11.2. Rationale for Nitroderm TTS efficiency

- These advantages result in better subjective well-being and quality of life for the patient, and greater satisfaction with the results on the part of the doctor.
- These therapeutic advantages bring about a reduction in the overall use of resources and in the loss of productive capacity.

International Study Program to Test These Hypotheses

A comprehensive international study program has been conducted in Europe, the United States, and Australia to provide evidence substantiating the product's value for money (Table 11.1). The program includes 14 studies conducted by independent and well-known experts in the field of medical and economic research. The studies cover all relevant socioeconomic effects related to drug therapy and reflect the state-of-the-art of socioeconomic evaluation of drug therapy.

Structure and Content of the Nitroderm TTS "Economic Dossier"

The Nitroderm TTS "Economic Dossier" presents the results of this study program in a standardized format. Economically relevant effects of the preparation are assigned to the categories discussed below (see also Fig. 11.3).

Table 11.1. The international study portfolio

Title/country/year	Author(s)	Study design	Main results
Cost-benefit analysis of the long-term prophylaxis of angina pectoris attacks with Nitroderm TTS, Switzerland 1982/83	HealthEcon Ltd., Basle	Open, multicenter, intraindividual comparison of two 3-month periods (before and after changeover to Nitroderm TTS)	Reduction in overall therapy costs by 23% in comparison with traditional therapies
Study of patient compliance with different forms of drug therapy for the prophylaxis of angina pectoris, Federal Republic of Germany, 1985/86	Prof. E. Weber, University of Heidelberg	Open, multicenter, prospective, intra-individual, randomized	Nitroderm TTS not only reduces the absolute difference between the prescribed medication and that actually taken but also makes more patients able or willing to follow the doctor's instructions
Cost-benefit analysis of the prophylactic medication of angina pectoris in the Federal Republic of Germany, 1986	Prof. P. Oberender, University of Bayreuth; Prof. B. König, University of Mainz; KFM-Clinical Research Inc., Munich	Open, multicenter, randomized, prospective, crossover, 2 × 3 months	Nitroderm TTS reduced overall costs of therapy by more than 6% in comparison with other angina pectoris medications
Cost-effectiveness analysis of angina pectoris prophylaxis with Nitroderm TTS in Austria, 1985/86	HealthEcon Ltd., Basle	Open, multicenter, retrospective, intraindividual comparison between two 3-months periods	Costs per patient per quarter were reduced by roughly 4%
Angina pectoris: prophylaxis and quality of life, Austria, 1986	HealthEcon Ltd., Basle	Open, randomized, prospective, multicenter, intraindividual comparison, control phase of 4 weeks	Nitroderm TTS brought about a marked improvement in quality of life aspects of angina pectoris patients and in this respect was superior to conventional therapy with long-acting oral nitrates
Efficacy and cost-effectiveness of angina pectoris prophylaxis as assessed by the Swiss medical profession, 1986	Dr. B. Horisberger, Interdisciplinary Research Center for Public Health, St. Gallen; HealthEcon Ltd., Basle	Representative personal survey	In the opinion of Swiss medical doctors, Nitroderm TTS offers advantages in the treatment of angina pectoris patients in comparison with long-acting oral nitrates
Quality of life of patients with angina pectoris, Sweden, 1986	Prof. B. Jönsson, University of Linköping, Center for Medical Technology Assessment;	Open, multicenter, prospective, crossover (2 × 4 weeks), blind randomization	When patients were divided according to preference, highly significant differences between Nitroderm TTS and long-acting oral nitrates were found (degree of relief, duration and frequency of cardiac pain, and general well-being)

Study title	Author	Method	Results
	Ass. Prof. S. Hovendahl, Helsingborg Hospital		
Pilot study for cost-benefit analysis of Nitroderm TTS in the treatment of patients with angina pectoris, United Kingdom, 1986	Dr. Chr. Bulpitt, Dr. A. Fletcher, London School of Hygiene and Tropical Medicine; Prof. A. Maynard, University of York	Open, retrospective, multi-center, between groups of patients	Overall, the cost for the transdermal treatment was higher than that for the oral treatment. However, there were more patients from NYHA classes 3 and 4 in the Nitroderm TTS group than in the control group
Therapy for angina pectoris with Nitroderm TTS, Federal Republic of Germany, 1983/84	Dr. H. Letzel, L. C. Johnson, Munich	Multicenter, open, change-over, prospective, 3-week observation period	37 596 patients with stable angina pectoris were observed. Markedly fewer angina pectoris attacks and less use of concomitant medication were reported
A study on patient satisfaction with Nitroderm TTS in Australia: life before and after the advent of Nitroderm TTS, 1986	Lenehan Lynton Bloom Blaxland, Sydney	Survey by self-administered questionnaires	75% of 113 patients stated that angina affects their life-style to a lesser degree after application of the patch
A study on the cost-effectiveness of Nitroderm TTS in the Michigan Medicaid Program, United States, 1986	Dr. G. Oster, Policy Analysis Inc., Boston; Dr. A. Epstein, Harvard School of Public Health, Boston; Prof. B. Weisbrod University of Wisconsin	Evaluation of computerized Medicaid data	Nitroderm TTS use did not result in less or more health care utilization
Transdermal nitrate therapy in coronary heart disease: the view of the patient, Federal Republic of Germany, 1986	Dr. R. Düsing, University Hospital, Bonn; Infratest, Munich	Survey of doctors in private practice and patients selected on a randomized basis	Nitroderm TTS improved handicap in housework, shopping, or professional work as well as social life-style. Results were statistically significant
Benefits derived from Nitroderm TTS therapy/quality of life, United Kingdom, 1986	Gallup Institute, London	Survey in patients by self-administered questionnaire	A diminution was observed in the extent to which angina attacks affected patients' hobbies and social life: from 80% before Nitroderm TTS therapy down to 57% during treatment
Nitroderm TTS Chart Review in private practice, United States, 1981–85	Dr. D. Chinoy, Jacksonville Cardiovascular Clinic, Florida	Review of 46 medical record files in a private cardiology practice	Nitroderm TTS led to less hospitalization and also decreased the average length of hospitalization

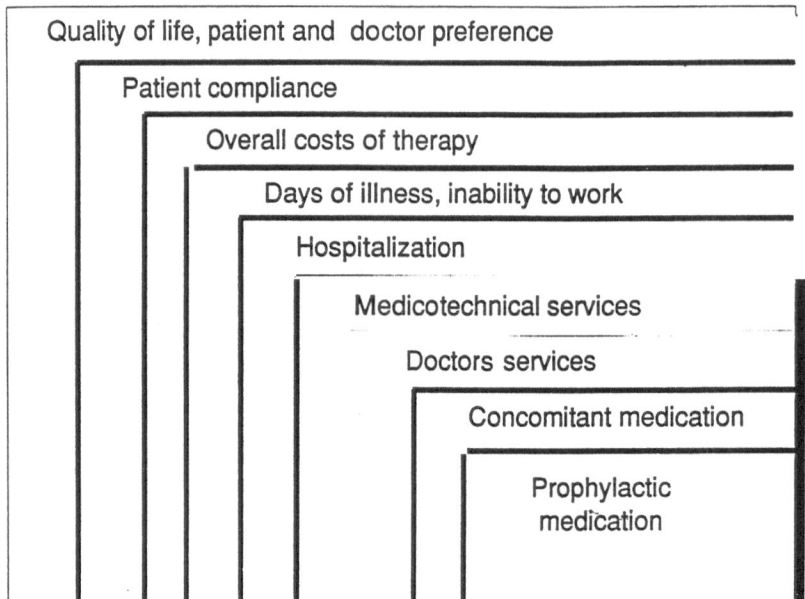

Fig. 11.3. Components of Nitroderm TTS efficiency

Medication

The treatment of angina pectoris pursues three aims:
- Prevention or reduction of attacks
- Terminating acute attacks as rapidly as possible when they occur
- Supportive or symptomatic measures, including the avoidance and/or treatment of adverse reactions

In line with these different therapeutic aims the drug therapy of angina pectoris includes three types of drugs:
- Medication for the prevention of attacks
- Medication for cutting short attacks (usually sublingual nitroglycerin)
- Medication for concomitant and adjuvant therapy (analgesics, tranquilizers, etc.)

The purpose of Nitroderm TTS is the prevention of attacks. However, the effects of its use are not limited to this area but also extend to the volume and costs of treating acute attacks and of adjuvant therapy.

A number of studies have been designed to determine the impact of Nitroderm TTS on additional causal, symptomatic, or side-effect induced medication. The studies indicate that the cost of medication solely for the purposes of attack prevention are higher with Nitroderm TTS than with traditional medication. The reason for this difference lies in the relatively higher daily costs of Nitroderm TTS. They also show, however, that the use of Nitroderm TTS reduces the amount and costs of medication for terminating acute attacks and for adjuvant therapy. This is a result of a reduction in the frequency and/or severity of attacks and the general improvement in the well-being of patients on Nitroderm TTS.

Doctors' Services

Angina pectoris patients require intensive medical and paramedical care. A number of studies have been designed to determine the impact of Nitroderm TTS therapy on the type, volume, and costs of these services. Some of the studies show that with use of Nitroderm TTS some doctors are able to limit the amount and/or the frequency of the services they provide. Presumably this is related to the effectiveness of Nitroderm TTS: the number and/or the frequency of attacks is reduced and the patient's general well-being improved, enabling the doctor to reduce the number of patient contacts.

Medicotechnical Services

Diagnosis of angina pectoris and the management of patients with this condition requires the use of medicotechnical procedures such as resting electrocardiograms, exercise electrocardiograms, blood tests (e. g., lipid spectrum, blood sugar), and even radiography. A number of studies have shown a reduction in the volume of these medicotechnical procedures for patients on Nitroderm TTS. This reduction is apparently related to the increased effectiveness of the adhesive patch in comparison to alternative methods of treatment and the consequent reduction in the amount of necessary medical care.

Hospitalization

In many cases angina pectoris necessitates the hospitalization of the patient. The Interdisciplinary Research Center for Public Health (IFZ)/Health Econ ascertained, for instance, that each year in Switzerland approximately 9300 patients (roughly 0.2% of the total population) with stable angina pectoris are treated in hospital.

Hospitals have long been a leading generator of health care costs. For this reason, even a small change in the number of hospital admissions and/or length of stay has a considerable impact on costs.

Days of Illness

The costs of an illness are not limited to the direct costs of medical and medicotechnical services, drugs, and stays in hospital. Illness also imposes an economic burden through work loss related to disease. This is particularly evident for patients engaged in a gainful occupation but also applies in principle to housewives. A reduction in the number of absences from work in persons of working age is therefore synonymous with a gain in economic productivity. The studies show that decreasing days off work is one of the decisive benefits of Nitroderm TTS therapy.

Overall Costs of Therapy

The term "overall costs of therapy" is defined variously in the individual studies. A number of studies limit themselves to an examination of the "direct costs," i.e., the costs of medication, medicotechnical services, and doctors' services. Other studies also include "indirect costs," as, for instance, in the form of lost working days and resultant losses of productivity. Detailed results of the studies vary due to differences in the scope of the studies and national differences in therapeutic regimes and in price and tariff structures. Most of the studies, however, show a reduction in overall costs in comparison with the alternative forms of therapy. Although Nitroderm TTS in all cases brings about an increase in medication costs for attack prophylaxis, this is compensated for by savings in other areas of therapy. It is therefore clear that in order to assess the cost-effectiveness of Nitroderm TTS, it is not enough to compare its per dose costs with those of alternative preparations. No verdict can be passed on the merits or demerits of Nitroderm TTS until the effects of its use in other cost areas of therapy have also been taken into consideration.

Patient Compliance

Patient compliance is a factor of steadily growing importance in connection with the cost-effectiveness analysis of a drug. Although patient compliance is not a valued end point in its own right, it leads to quantifiable benefits in other areas.

Since it does not follow automatically that the patient will comply with the doctor's instructions, the consequences of the patient's unwillingness to cooperate in his treatment are the subject of widespread discussion. From the medical viewpoint, failure to take prescribed medication can endanger the success of treatment and lead to complications. From the economic viewpoint, drugs which are prescribed and received but not used are a waste: It is estimated that in the Federal Republic of Germany alone, today the economic loss due to drugs that are not taken amounts to 2–3 billion DM.

An endeavor to counteract this development is made by the use of 'patient- and treatment-friendly' forms of drugs. Nitroderm TTS is an example of this. Its use allows therapeutic regimes to be simplified and dosage errors to be reduced, and encourages the patient to take the medication as instructed. A study specially conducted with this objective shows that these are important steps toward an improvement in patient compliance.

Quality of Life

Since the introduction of Nitroderm TTS, it has been used daily by a steadily growing number of patients: at present nearly one million all over the world. As many of these patients are paying the drug bill out of their own pocket, this "willingness to pay" suggests that they prefer the patch, underlining the improved quality of life that is achieved. Publications have also drawn attention to the high patient acceptance of Nitroderm TTS. The preparation has obviously brought special advantages and benefits from the viewpoint of specific patients, who feel that their quality of life is improved as a result.

Changes in the quality of life are an expression of the "benefit" which the patient obtains from treatment in a favorable case, or the "costs" he must bear in an unfavorable case. Cost-effectiveness analyses of medical care are incomplete if they do not take into account these aspects which are of decisive importance for the patient. A conclusive judgement on the value of a form of treatment can be reached only by considering the objective clinical parameters of the doctor, the monetary parameters of the economist, and the subjective parameters of the patient. Several studies have been specially designed to examine the effects of Nitroderm TTS on the subjective well-being of angina patients, their objective life circumstances, and their treatment preference.

Conclusions

The individual studies should not be seen in isolation; all the studies taken together form the total picture of the economic performance of Nitroderm TTS. Today it is among the best documented of pharmaceutical products as far as cost-effectiveness is concerned. In summary, taking all studies into account, Nitroderm TTS is an economic alternative to existing therapies on the drug market. Ciba-Geigy accepts the changing environment and the need to demonstrate the cost-effectiveness of its products. Nitroderm TTS has been the first of its products on which an international research program has been carried out. In future Ciba-Geigy will undertake cost-effectiveness research for all its main products. Investigations in this direction will not start only once the product has entered the market place; rather economic considerations will already play a role during the product development process.

12. Cost – Cost Comparisons of Nitroderm TTS in Switzerland and Austria

R. H. Dinkel

Institutional Background

As part of the ongoing efforts in Switzerland and Austria to stem rising costs in health care services, particular attention has been devoted to the role played by drugs. In both countries, powers have been established for limiting the number of reimbursable drugs by drawing up formularies (*Spezialitätenliste* in Switzerland, *Heilmittelverzeichnis* in Austria). One of the decisive criteria for admitting a new preparation to these lists is cost-effectiveness [3, 5]. A manufacturer has to demonstrate to the authorities that the price of his new preparation is justifiable when compared to preparations already on the list. It is no longer sufficient to prove efficacy and safety to get full approval for reimbursement. Nowadays one also has to prove its economic merits, its efficiency.

Study Objective

The demand for rational, cost-effective prescription of drugs is undoubtedly justified. Not everything can or should be financed. This demand must also be satisfied by innovative products such as the transdermal therapeutic system. Nitroderm TTS has been described as expensive because the costs per pack and per dose are higher than those of traditional nitrate preparations in tablet, capsule, or ointment form. This price difference can be explained as a result of the research, development, and manufacturing costs. Higher unit costs, however, can be justified only if the preparation gives comparatively greater benefits or cost savings in other areas of angina pectoris therapy. Evidence must be provided that, despite the per dose price difference, transdermal therapy fulfills the overall cost-effectiveness demand of giving value for money.

Two studies are presented here, one conducted in Switzerland in 1983 [1], the other in Austria in 1986 [2]. The primary objective of these studies was to determine whether the use of Nitroderm TTS represents a reasonable alternative, in terms of cost-effectiveness, to the preparations traditionally used in these countries for the prophylaxis of angina pectoris attacks.

Cost-Effectiveness Parameters

Cost-effectiveness assessment of a drug can take into consideration a wide spectrum of relevant drug effects. This spectrum ranges from impact on the use of concomitant medication, doctors' services, and medicotechnical procedures, through effects on the number of ambulatory or hospitalized days of illness and absence from work, to qualitative influences on compliance and the mental, physical, and social well-being of the patient.

The two studies presented have limited their scope to the so-called direct costs, i. e., those which are measurable in monetary units. In terms of cost-effectiveness terminology these studies consequently can be labeled as cost–cost comparisons [4].

Both studies evaluated the following cost elements:
- Overall antianginal *medication*, whether for prevention of angina attacks, for cutting short attacks, or for supportive, causal, or symptomatic treatment, including therapy for unwanted side-effects of medication
- *Services* provided by physicians, in the context of antianginal therapy, whether in the form of house visits or of consultations (either in the practice or by telephone)
- *Medicotechnical procedures* required for the diagnosis of angina pectoris and management of patients, such as resting or exercise electrocardiograms, blood tests, and radiography

Study Design

These parameters were investigated in open, retrospective, multicenter studies carried out on ambulatory patients. The subjects consisted of 25 men and women in Switzerland and 63 men and women in Austria who had been switched by their physicians from conventional medication to Nitroderm TTS. Comparison was made between two 3-month periods, one before changeover to Nitroderm TTS and one thereafter.

Relevant data were obtained using a structured questionnaire which was filled in by 12 doctors in Switzerland and 21 doctors in Austria on the basis of their patient files. The patients themselves were not questioned.

Switzerland, 1983

Medication

To determine the effects of Nitroderm TTS therapy on overall antianginal medication the preparations were classified into the following groups (Table 12.1):
- Medication for prevention of attacks
- Medication for cutting short attacks
- Medication for adjuvant therapy (for treatment of causes, symptoms, or side-effects)

Table 12.1. Costs of medication in Swiss study (average monthly costs in SFr per patient)

Groups of preparation	Before change of therapy to Nitroderm TTS	After change of therapy to Nitroderm TTS	Difference	
			Abs.	%
Prevention of attacks	37.80	42.00	+ 4.20	+ 11.1
Cutting short attacks	10.50	2.40	− 8.10	− 77.1
Adjuvant therapy	46.20	31.80	− 14.40	− 31.2
Total	94.50	76.20	− 18.30	− 19.4

Attack prevention medication cost an average of SFr. 37.80 per month before the changeover. After the changeover, this sum increased to SFr. 42.00. This corresponds to an absolute increase of SFr. 4.20 or around 11%. This increase was due mainly to the higher daily dosage costs of Nitroderm TTS compared with those of the previous treatment.

On the other hand, with regard to the therapeutic aim of *cutting short attacks,* on Nitroderm TTS the average cost of drugs per patient per month was reduced by more than 77%, from SFr. 10.50 to SFr. 2.40. This reduction was not due to a change in the *type* of preparations given for the treatment of attacks; these were largely identical to those given before the changeover. The reason for the reduction was a decrease in the *amount* of medication given, due to a smaller number of angina pectoris attacks among patients on the transdermal treatment.

There was also a substantial reduction in medication used for *adjuvant therapy.* Before the changeover, the cost of these medications (primarily beta-blockers, calcium antagonists, digitalis, and tranquilizers) averaged SFr. 46.20 per month. After the changeover to Nitroderm TTS, the monthly cost of concomitant therapy amounted to SFr. 31.80, a reduction of approximately 31%. The reason for this saving was a lowering of the dosage required or cessation of other medications. This was, again, presumably a result of reduced frequency and/or severity of attacks and a general improvement in well-being on Nitroderm TTS.

Although assessment of the costs of prophylactic medication alone showed that Nitroderm TTS brought a cost disadvantage, a more comprehensive analysis showed that Nitroderm TTS therapy had a substantial cost advantage. After the changeover, an average of SFr. 76.20 per month was spent on drugs compared with SFr. 94.50 before the changeover, a reduction of nearly 20%.

Doctors' Services

The average cost of doctor consultations was SFr. 35.33 per month before the changeover to Nitroderm TTS. After the changeover, this cost was reduced to SFr. 23.33, or by around 34% (Table 12.2).

Table 12.2. Costs of services in Swiss study (average monthly costs in SFr per patient)

Nature of services	Before change of therapy to Nitroderm TTS	After change of therapy to Nitroderm TTS	Difference	
			Abs.	%
Consultations	35.33	23.33	− 12.00	− 34.0
Medicotechnical services	10.09	8.15	− 1.94	− 19.2
Total	45.42	31.48	− 13.94	− 30.7

The cause of this reduction was an improvement in patients' clinical symptoms which allowed their doctors to lengthen the interval between consultations. Prior to the changeover, patients consulted their doctors an average of 4.5 times per quarter, compared with only 3.3 times afterwards.

Medicotechnical Services

The study also showed a reduction in costs due to medicotechnical services (Table 12.2). Before the changeover to Nitroderm TTS, an average of SFr. 10.09 per patient per month was spent specifically on electrocardiograms and exercise tests. After the changeover, the relevant sum amounted to SFr. 8.15, which corresponds to a cost reduction of approximately 19%. This reduction was almost solely due to the need for fewer electrocardiograms. Relatively few exercise tests were conducted, perhaps as a result of the relatively high average age of the patients.

Overall Costs of Therapy

The results of the Swiss study showed that the changeover to Nitroderm TTS increased the costs of attack prophylaxis medication by around 11%. However, this increase was compensated by savings in other areas.

Treatment of angina pectoris before the changeover to transdermal therapy resulted in average monthly costs of SFr. 139.92 per patient (Table 12.3). After the

Table 12.3. Overall costs in Swiss study (average monthly costs in SFr per patient)

Cost category	Before change of therapy to Nitroderm TTS	After change of therapy to Nitroderm TTS	Difference	
			Abs.	%
Medication	94.50	76.20	− 18.30	− 19.4
Services	45.42	31.48	− 13.94	− 30.7
Overall costs	139.92	107.68	− 32.24	− 23.0

changeover, this sum amounted to SFr. 107.68. Thus the use of Nitroderm TTS reduced costs by 23% in comparison with traditional therapies.

Austria, 1986

The Austrian study had similar objectives to the Swiss study, but in addition addressed the question of whether the Swiss results are reproducible in a different therapeutic and economic environment. A much larger sample size, 63 patients, was studied in Austria.

Medication

Like the Swiss study, the Austrian study showed a substantial increase in medication costs for *attack prophylaxis* (Fig. 12.1). Before the changeover to Nitroderm TTS, average medication costs for this therapeutic aim were S 758.36 per patient per quarter. After the changeover, these costs increased by approximately 85% to S 1422.32. This increase was due, on the one hand, to the higher daily costs of Nitroderm TTS and, on the other, to the fact that many doctors continued prescribing oral or dermal preparations in addition to the new product.

Sublingual nitroglycerin is the agent of choice for *cutting short* angina attacks. Before the changeover to Nitroderm TTS, S 294.01 per patient per quarter was spent on these drugs; after the changeover, the figure was only S 60.59, corresponding to a cost reduction of S 233.40 or 79.4%.

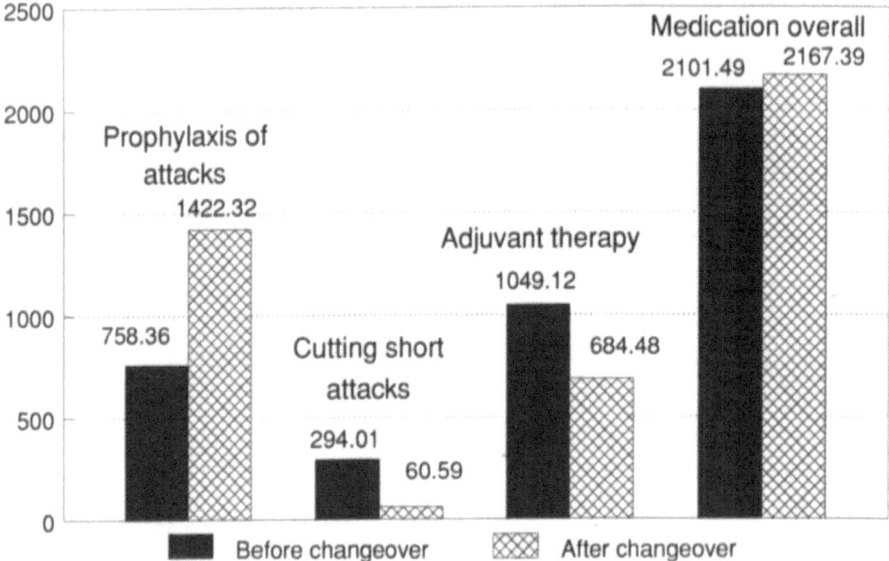

Fig. 12.1. Costs of medication in the Austrian study (S per patient/quarter) before and after changeover to Nitroderm TTS

According to the physicians in the study, the main reason for this reduction was a decrease in the frequency of attacks among patients treated with Nitroderm TTS. The average number of attacks diminished approximately 80% after the changeover. The duration and severity (painfulness) of the attacks was also reduced.

A substantial reduction in drug costs also took place for medications used as *adjuvant therapy*. Prior to the changeover to Nitroderm TTS, the cost of adjuvant preparations averaged S 1049.12 per patient per quarter, whereas after the changeover it was reduced to S 684.48, a decrease of S 364.64 or 34.8%. This drop was due to the discontinuation of individual products or a reduction in dosage, made possible both by diminution in frequency and severity of attacks and by reduction of side-effects.

The average *total expenditure on medication* before the changeover to Nitroderm TTS was S 2101.49 per patient per quarter, compared with S 2167.39 after the changeover. This corresponds to an increase of S 65.90, only 3.1% in terms of the initial expenditure.

Doctors' Services

A comparison was also made between doctors' services provided before and after the changeover. The doctors' services were classified as follows:
– Scheduled and spontaneous visits
– Home visits
– Telephone contacts

As a result of the changeover to Nitroderm TTS, the doctors were able to reduce the volume of these services substantially (Fig. 12.2). Before the changeover there were

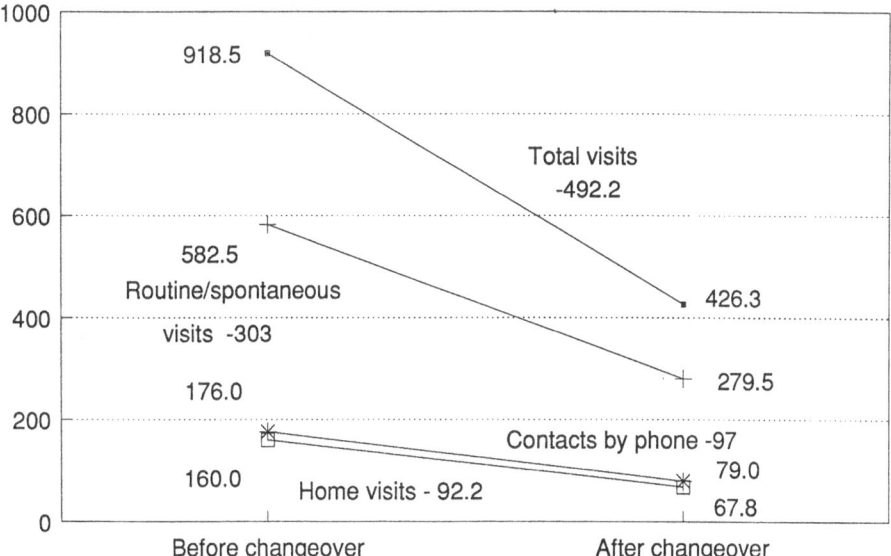

Fig. 12.2. Consultations in the Austrian study (total per quarter) before and after changeover to Nitroderm TTS

Table 12.4. Medicotechnical services in Austrian study (total per quarter) before and after changeover to Nitroderm TTS

Type of procedure	Number		Decrease
	Before changeover	After changeover	on Nitroderm TTS
Electrocardiograms	125	74	− 51
Laboratory tests	75.75	51	− 24.75
Intravascular injections	48.5	2	− 46.5
Neural therapies	89	12	− 77
Exercise tests	7.25	1.25	− 6
Venesections (blood tests)	3	–	− 3
Chest X-rays	1.5	1	− 0.5
Ultrasonic treatments	2.5	1.25	− 1.25
Thallium scintigraphy	1	–	− 1
Oxygen treatments	2	–	− 2

roughly 14 doctor–patient contacts of various types during the quarter; after the changeover there were only appoximately seven contacts. Considering patients individually, consultations diminished in 51 cases, were unchanged in eight, and increased in four.

In Austria, it is difficult to determine the cost savings from consultation reductions, since these are covered by an inclusive insurance system. Only home visits are paid for separately as special services. Based on a charge rate of S 111.00 per visit, the cost of home visits before the changeover averaged S 281.90 versus S 119.46 after the changeover, a reduction of more than 57%.

Medicotechnical Services

The analysis included all medicotechnical procedures employed for the diagnostic confirmation of angina pectoris or for its follow-up and therapy, whether performed by the patients' doctors or elsewhere. Comparison of the 3-month periods before and after the changeover to Nitroderm TTS showed a marked fall in the volume of such services (Table 12.4).

Overall Costs of Therapy

In the overall examiniation of *medication* and *home visits*, the group of patients included in the study showed a reduction in costs on Nitroderm TTS compared with the situation prior to the changeover (Table 12.5).

Table 12.5. Overall costs in Austrian study (average costs in S per patient and quarter)

Cost category	Before change of therapy to Nitroderm TTS	After change of therapy to Nitroderm TTS	Difference	
			Abs.	%
Medication	2101.49	2167.39	+ 65.90	+ 3.1
Home visits	281.90	119.46	− 162.44	− 57.6
Overall costs	2383.39	2286.85	− 96.54	− 4.1

After the changeover a total of S 2286.85 was spent per patient per quarter, compared with S 2383.39 before hand. Thus, the use of Nitroderm TTS reduced the cost per patient per quarter by S 96.54 or roughly 4.1%

Conclusions

Several limitations of these studies must be mentioned. The studies under review used a relatively simple methodological approach. Also, the patient population was comparatively small. The predominant purpose of the studies, therefore, is an *exploratory* one; their results must yet be proved by trials based on a representative number of patients and designed according to more rigorous criteria. Despite these reservations, the following methodological and factual conclusions may (tentatively) be drawn from the results:

– Both studies used the same design and the same study parameters, limited to an analysis of direct costs. Yet the numerical results of the studies with respect to costs and savings differ considerably. This is mainly due to national differences in therapeutic regimes, reimbursement systems, tariff structures for services, and price structures for goods. The results of cost-effectiveness oriented studies, therefore, cannot automatically be transferred to other countries. The validity of cost – cost comparisons is particularly limited to specific countries and circumstances.

– Regarding its cost-effectiveness, Nitroderm TTS therapy has various effects. Considering attack prevention alone, the use of Nitroderm TTS leads to an increase in drug costs compared with the conventional alternatives. However, if the consideration is extended to include other cost components of angina pectoris therapy, Nitroderm TTS also brings reductions. These reductions are of a large enough magnitude that, taken as a whole, they offset the increased costs related to prophylaxis. As a result, the overall costs of angina pectoris therapy are reduced.

– This result illustrates how just a straightforward comparison of dosage costs is an inadequate, if not misleading, measure to determine the cost-effectiveness of drugs. Medicines which, on the face of it, are more expensive can be more cost-effective than cheaper preparations if they lower the cost of other aspects of therapy and can make for a better overall cost–benefit ratio. In these cases, a meaningful cost-effectiveness assessment of drugs can only be made if all economically relevant effects of using the preparation are taken into account as comprehensively as possible.

References

1. Dinkel R (1983) Ökonomische Analyse eines neuen Therapiekonzeptes: Nitroderm TTS. Swiss Pharma 10:27–29
2. Dinkel R (1986) Die transdermale Therapie der Angina pectoris im Wirtschaftlichkeitstest. Österreichische Ärztezeitung 19:54–56
3. Imbach S (1986) Aufnahme von Arzneimitteln in die Spezialitätenliste. Schweizerische Zeitschrift für Sozialversicherung und berufliche Vorsorge 1:1–26
4. Little AD (1984) Cost-effectiveness of pharmaceuticals: report. Pharmaceutical Manufacturers Association, Washington, D.C.
5. Nord D (1982) Modell Österreich?: Gesundheitssystem und Arzneimittelversorgung in Österreich. Med.-Pharmazeut. Studiengesellschaft, Mainz

13. The Cost-Effectiveness of Transdermal Nitroglycerin in the Michigan Medicaid Program: A Preliminary Report

D. M. HUSE, G. OSTER, B. A. WEISBROD,
J. L. READ, B. L. LUCE, and A. M. EPSTEIN

Introduction

Nitroglycerin and related nitrate agents are the most widely prescribed medications for angina pectoris, a chronic manifestation of coronary heart disease (CHD) that affects approximately 2 million persons in the United States [1]. While some nitroglycerin products are used to treat acute anginal attacks, others – notably, long-acting oral nitrates (LAOs) and transdermal nitroglycerin (TDN) patches – are used on a daily basis to prevent such attacks.

Transdermal nitroglycerin, which provides rate-controlled topical administration of nitroglycerin, requires less frequent dosing than LAOs, which must be taken 2–4 times daily. Patients and physicians have generally expressed a high degree of satisfaction with TDN; the extent of such satisfaction is perhaps suggested by the degree to which it has penetrated the U.S. antianginal market. Despite costing more than LAOs, TDN currently accounts for about 30% of sales of all antianginal medications (unpublished data, IMS America Ltd.).

The higher daily cost of TDN, however, has led some to question its cost-effectiveness. It is well known, though, that comparison of drug costs alone may give rise to mistaken conclusions regarding the cost-effectiveness of alternative therapeutic strategies. A more costly drug may also be more efficacious, and may reduce the need for other components of care. Thus, in the present instance, patients' satisfaction with TDN may be reflected in fewer visits to physicians' offices or hospital emergency rooms, or in the reduced use of concomitant medications. Such benefits of TDN, if they exist, could offset the higher daily cost of therapy.

In this study, therefore, we examined the cost-effectiveness of TDN in the treatment of angina pectoris, using patient-level claims data from the Medicaid program in the state of Michigan. The question we posed is whether there is any difference in overall levels of health care utilization and cost between patients receiving TDN and those receiving LAOs.

Data Sources

To examine the cost-effectiveness of TDN, it was necessary to obtain a complete record of health care utilization for a large number of patients who had been treated with either TDN or LAOs. The Medicaid program in the state of Michigan, which pays for a comprehensive range of medical goods and services and which requires minimal patient copayments, represented such a source of utilization data. Consequently, data from this program were obtained for the years 1982 and 1983, the most recent available at the time the study was initiated.

Enrollment information and all health care provider claims for each beneficiary were included in these records. Enrollment data included each beneficiary's age, sex, race, and county of residence. They also indicated, for each calendar month, whether that individual was currently eligible for Medicaid benefits, and whether he or she had any additional health insurance coverage. Provider claims included those for hospitalization, physician visits, other outpatient services, long-term care, home health care, and drugs. With each claim was recorded the date of service, the amount charged, and the amount reimbursed by Medicaid. Drug claims indicated the specific agent purchased, while nondrug claims included at least one ICD-9-CM diagnosis code. Hospitalization claims also included a second diagnosis code, a procedure code (for example, indicating surgery or diagnostic workup), and the length of stay.

To select samples of TDN- and LAO-treated patients, all drug claims for TDN or LAOs and all nondrug claims with a diagnosis of angina pectoris were drawn from the 1982–1983 Michigan Medicaid data base, and the beneficiaries identified. Initially, all patients were selected who had at least one claim for TDN or LAOs *and* at least one recorded diagnosis of angina pectoris. To improve the reliability of the data, individuals less than 30 years of age were excluded.

Enrollment data and all claims for nondental health care goods and services during 1982 and 1983 were then compiled for the selected patients, and each person's claims were summarized by calendar month. A patient's utilization in any given month was measured only if that individual was enrolled in Medicaid during that month and was not simultaneously covered by Medicare or any other type of health insurance. This ensured that the data that were analyzed represented a complete record of health care utilization for each patient over the period of study. The resulting data base included, for each selected patient, demographic and diagnostic data and up to 24 monthly sets of utilization measures. As detailed below, we included in the study only those patients for whom we had at least 18 continuous months of utilization data.

Methods

The principal methodological issue in any retrospective evaluation study is the likelihood that the effect of treatment (i.e., TDN versus LAOs) on outcomes is confounded with the effects of other characteristics on which treatment and control groups differ. In a true experiment, random assignment of patients to treatment groups minimizes this risk, by ensuring that the two groups of patients do not differ in any systematic fashion. The use of nonexperimental data, such as Medicaid claims, instead requires careful study design and the use of appropriate statistical methods to control for between-group differences that might bias any comparison of outcomes.

Selection of Treatment and Control Groups

An important consideration in the design of an evaluation study is the selection of comparable treatment and control groups. Patients observed (by their health care claims) to have received TDN or LAOs might, for example, have been starting a new course of therapy, they might have been on continuous therapy for some time, or they might have changed from LAO to TDN therapy (or vice versa). Recognizing the various patterns of use of one or both medications among Michigan Medicaid benefi- ciaries, we initially considered several possible options for defining a *treatment group of TDN users* and a *control group of LAO users*. Each alternative reflected a specific question of clinical and economic interest:

1. Do levels of health care utilization differ between patients starting therapy with TDN and those starting with LAOs?
2. Do levels of health care utilization differ between patients switching from LAOs to TDN and those switching from TDN to LAOs?
3. Do levels of health care utilization differ between patients switching from LAOs to TDN and those remaining on LAOs?
4. Do levels of health care utilization differ between patients switching from TDN to LAOs and those remaining on TDN?
5. Do levels of health care utilization differ between patients on long-term therapy with TDN and those on long-term LAO therapy?

The pursuit of options 2 and 4 would have required that the Michigan Medicaid data base contain a sizable number of patients who had used TDN for some time before switching to LAOs. Since TDN was approved for use in the United States in 1982, very few patients would have been identified for a TDN-to-LAO "switchers" group.

These two options were therefore ruled out as impractical with the available data. Option 5 was also considered impractical because it would have required a group of TDN-treated patients on long-term therapy in 1982–1983. Moreover, this option was precluded by the pretest–posttest design (described below) that we sought to employ. Finally, options 3 and 4 were considered to be seriously flawed due to their inherent dissimilarity between treatment and control groups. Presumably, patients who switched therapies did so because they were not responding adequately, and thus the switchers, on average, were probably more severely ill than the nonswitchers. On the other hand, the immediate result of any change in therapy, even to a placebo, tends to be an improvement in health status, which in turn may be reflected in lower utilization [2]. This "Hawthorne effect" would bias estimates of cost-effectiveness in favor of the therapy to which patients had switched.

To ensure a sufficiently large sample of roughly similar TDN and LAO patients, a "new starts" design (option 1) was employed. *New starts* were defined as patients who began therapy with TDN or LAOs following a period of at least 6 months during which neither agent was used. To be included in the sample, therefore, the health care utilization of each TDN- or LAO-treated patient had to be observed for at least 6 months prior to the first recorded claim for TDN or an LAO. Patients with claims for both agents were assigned to treatment or control groups on the basis of the first such claim. Patients who were started on both medications in the same month were excluded from the analysis.

Subgroups of TDN and LAO new starts were also defined for separate analysis. One such group, termed *nonswitchers*, consisted only of those new starts who did not have claims for both TDN and LAOs. Another subgroup, called *continuers*, consisted of those nonswitchers who continued to have claims for TDN or LAOs in the 12th month after initiation of therapy. The advantage of these more restrictive sample definitions is that they exclude changes or lapses in therapy, the effects of which would otherwise be attributed to the initial therapy received by the patient. Their disadvantage is that they obscure any differences between TDN and LAO therapies in patients' compliance or physicians' satisfaction, as indicated by continued prescription.

Pretest–Posttest Design

The research design employed in this study involved the measurement of each patient's health care utilization both before and during treatment. In the language of evaluation research, we used a *pretest–posttest design* [3]. The outcome that was compared between TDN- and LAO-treated patients was their level of health care utilization during treatment. The measurement of pretreatment utilization provided a means of controlling for differences in outcome that were unrelated to treatment.

It is possible, for example, that patients who received the newly approved TDN in 1982 were sicker on average than those who were prescribed LAOs. TDN-treated patients may therefore have required more intensive health care, creating the appearance that TDN causes higher levels of utilization – even though the opposite could be the case. Alternatively physicians who were among the first to learn of and prescribe TDN may have been specialists in cardiovascular medicine, who pursued more aggressive and costly treatment strategies than their LAO-prescribing peers. Or it may be the case that, for unknown reasons, one of these groups of patients contained significantly more individuals who were "high utilizers." The inclusion of pretreatment measures of utilization in the analysis was designed to permit such confounding factors to be detected, and thereby separated from the effect of treatment per se.

After examining the numbers of TDN- and LAO-treated patients having different numbers of months of continuously observed utilization data, pretreatment and treatment periods of 6 and 12 months, respectively, were specified. A follow-up period of 12 months allowed ample time to assess any effect of treatment on utilization, while a total observation period of only 18 months out of a possible 24 permitted sufficient sample size.

Examination of monthly utilization levels over the 18-month period of observation revealed a characteristic peak around the month of the first claim for TDN or LAOs. The initiation of therapy was often associated with an episode of hospitalization, which we considered likely not to be an outcome of treatment, but rather an independent factor affecting that outcome. We therefore established a third study period, consisting of the 1st month of treatment and the prior month. Utilization measures from this period were used to improve control for differences among patients–such as disease severity – that might have confounded the effect of treatment. This modification, however, did not fundamentally alter the pretest – posttest framework of the study design.

The design that was employed, therefore, specified three periods for each TDN- or LAO-treated patient. *Period 1* (the pretreatment period) included the 6th through the 2nd months prior to the start of treatment. *Period 2* (the initiation period) comprised the month prior to the start of treatment and the 1st month of treatment. *Period 3* (the treatment period) included the 2nd through the 12th months of treatment – the period during which the outcome of treatment was measured.

Model of Utilization

To examine the effect of TDN on health care utilization, it was first necessary to have a model of all the factors, including choice of therapy (i.e., TDN or LAOs), that determined treatment-period levels of utilization among the patients being studied. The basic model that we employed was as follows:

$U_3 = f U_1, U_2, P, T, Tx)$;

where

U_3 = period 3 utilization (during treatment);

and

U_1 = period 1 utilization (before treatment);
U_2 = period 2 utilization (initiation of treatment);
P = patient characteristics;
T = temporal factors; and
Tx = choice of therapy (i.e., TDN or LAOs).

This model was estimated using ordinary least squares (OLS) regression techniques. The coefficient of the choice of therapy variable provided a direct estimate of the effect of TDN use on utilization during period 3 – that is, of the cost-effectiveness of TDN versus LAOs. A positive coefficient would suggest that TDN therapy is associated with higher overall levels of health care utilization; a negative coefficient, on the other hand, would suggest that it is associated with lower levels.

Dependent variables included total health care charges, hospital charges, physician charges, and numbers of days hospitalized with a diagnosis of CHD. These were all measured as the natural logarithm of average monthly utilization during period 3; hence the estimated coefficients represent proportionate changes in the dependent variable. The explanatory variables that were employed in all regressions are presented in Table 13.1.

An alternative model was also employed to control more effectively for preexisting differences between the TDN and LAO patient groups. It has been demonstrated that, when assignment to treatment is nonrandom (as is generally the case in retrospective analyses), unbiased estimates of the effect of treatment can be obtained using a two-stage regression technique [4]. In this approach, the treatment variable (i.e., choice of therapy) is replaced with predicted values, which are derived from a prior regression of that variable on the remaining explanatory variables. The two-stage model that we employed was as follows:

$\hat{T}x = f(U_1, U_2, P, T)$; and
$U_3 = f(U_1, U_2, P, T, \hat{T}x)$;

where

$\hat{T}x$ = expected choice of therapy (i.e., TDN or LAOs)

Table 13.1. Explanatory variables in the model of utilization

---

Period 1 utilization:

Total health care charges
Numbers of drug claims
Any hospitalization with diagnosis of CHD (1 if yes, 0 if no)
Any hospitalization with other diagnosis (1 if yes, 0 if no)
Any claims for selected cardiac medications (1 if yes, 0 if no)

Period 2 utiliziation:

Total health care charges
Numbers of drug claims
Any hospitalization with diagnosis of CHD (1 if yes, 0 if no)
Any hospitalization with other diagnosis (1 if yes, 0 if no)

Patient characteristics:

Age, sex, and race
County of residence
Diagnosis of hypertension and/or heart failure

Temporal factors:

Month of first claim for TDN or LAO

Choice of therapy:

Treatment group (1 if TDN, 0 if LAO)

---

and all other indicated variables, dependent and explanatory, are unchanged from the single-stage model.

The first stage of this model was estimated with a logistic regression, and the resulting coefficients were used to calculate tne predicted treatment group for each patient. The second stage of the model was then estimated, as in the single-stage version, using OLS regression techniques.

Results

A total of 771 patients were included in the study, of whom 149 were TDN new starts and 622 were LAO new starts. Comparisons of these groups, presented in Table 13.2, reveal substantial differences. TDN-treated patients were slightly older, more likely to be white, and more often from a rural area or urban area other than Detroit. They were more likely to have a diagnosis of heart failure and used more cardiovascular medication, such as diuretics, beta-blockers, calcium channel blockers, and nitrates other than TDN or LAOs. They also required more hospital care, especially for the treatment of CHD. On the whole, therefore, TDN-treated patients had a different demographic makeup and were more severely ill (prior to starting TDN therapy) than LAO-treated patients.

Table 13.2. Comparison of TDN and LAO users: all new starts with 12-month minimum treatment period

Variable	TDN ($n = 149$)	LAO ($n = 622$)	P
Demographic characteristics			
Age (in years)	50.6	49.0	0.048
Sex (% male)	28.2	27.7	NS
Race (% white)	55.7	42.3	0.003
Residence (%)			
Metropolitan Detroit	35.6	71.5	
Other urban	44.3	19.2	< 0.001
Rural	20.1	9.3	
Comorbidities			
Hypertension (% diagnosed)	68.5	74.1	NS
Heart failure (% diagnosed)	53.0	38.1	< 0.001
Pretreatment medication			
Other Nitrates (% using)	30.2	19.5	0.004
Calcium channel blockers (% using)	14.1	5.6	< 0.001
Beta-blockers (% using)	29.5	20.9	0.024
Antihypertensives/diuretics (% using)	63.1	53.7	0.038

Comparison of Utilization Before and During Treatment

In Table 13.3 are displayed mean monthly levels of utilization, by study period, for TDN- and LAO-treated patients. Patients who received TDN had higher levels of health care utilization, variously measured, both before (period 1) and after the initiation of treatment (period 3). In the case of monthly drug charges, the gap widened during period 3, presumably reflecting the higher cost of TDN compared with LAOs. Utilization profiles for both groups also reveal a characteristic peak during period 2, suggesting that the initiation of TDN or LAO therapy was often associated with an episode of relatively intensive health care. TDN-treated patients, in particular, were more likely to have experienced a CHD-related episode of hospitalization during this period.

Of special interest is a comparison of the *change* in levels of utilization from period 1 to period 3 for TDN versus LAO users. While TDN-treated patients, as expected, experienced a much larger increase in monthly drug charges, their *total* monthly health care charges remained essentially constant (excluding the "spike" during period 2), implying that nondrug charges declined during period 3. LAO-treated patients, on the other hand, experienced a definite increase in utilization between periods 1 and 3. These results suggest that treatment with TDN may indeed be

Table 13.3. Comparison of TDN and LAO users: monthly utilization during periods 1, 2, and 3

Variable and period	Mean (std. error)		
	TDN	LAO	P
Total health care charges:			
Period 1	$ 860 (115)	$ 426 (35)	< 0.001
Period 2	2127 (249)	1173 (79)	< 0.001
Period 3	859 (77)	633 (40)	0.012
Hospital charges:			
Period 1	$ 559 (97)	$ 225 (28)	0.001
Period 2	1530 (208)	745 (64)	< 0.001
Period 3	489 (56)	344 (32)	0.041
Physician charges:			
Period 1	$ 179 (20)	$ 106 (7)	0.001
Period 2	399 (48)	214 (18)	0.016
Period 3	212 (19)	157 (8)	0.008
Drug charges:			
Period 1	$ 37 (3)	$ 25 (1)	< 0.001
Period 2	63 (3)	41 (1)	< 0.001
Period 3	63 (4)	42 (1)	< 0.001
Days in hospital with diagnosis of CHD:			
Period 1	0.16 (.06)	0.10 (0.03)	NS
Period 2	1.58 (.23)	0.61 (0.08)	< 0.001
Period 3	0.47 (.24)	0.12 (0.01)	NS

associated with reduced overall levels of utilization, relative to treatment with LAO s – especially for nonpharmalogical goods and services. These initial findings, however, lack the analytical reigor provided by multivariate regression analysis.

Regression Analysis

The coefficients of the choice of therapy variable, estimated using the single-stage model, are presented in Table 13.4. Positive coefficients indicate association of TDN therapy with higher levels of utilization. These coefficients are roughly interpretable as proportionate differences between TDN- and LAO-treated patients in the dependent variable. (With a logarithmic dependent variable and a continuous explanatory

Table 13.4. OLS regression coefficients for choice of therapy variable

Dependent variable (period 3 utilization)	All new starts		Nonswitchers		Continuers	
	Coeff.	(P)	Coeff.	(P)	Coeff.	(P)
Hospital charges	0.435	(0.337)	0.572	(0.263)	2.207	(0.018)
Physician charges	0.025	(0.891)	−0.034	(0.875)	0.300	(0.471)
Total charges	0.176	(0.210)	0.205	(0.204)	0.563	(0.032)
Hospital days with CHD	0.215	(0.021)	0.256	(0.007)	0.464	(0.004)

variable, this interpretation is exact; with a binary explanatory variable it is not.) The estimated effect of treatment among all new starts was generally positive, although usually not statistically significant. Only hospitalization with a diagnosis of CHD was significantly greater among TDN-treated patients. These findings are obviously at variance with those of the univariate analysis.

The single-stage model was also estimated with the "nonswitcher" and "continuer" subsamples. The results for nonswitchers were quite similar to those for all new starts. TDN use was associated with increased hospitalization for CHD, while other utilization measures were not significantly different between TDN and LAO nonswitchers. Among continuers, however, TDN was associated with significantly more days of hospitalization with diagnosis of CHD and higher total health care charges. Only physician charges were not significantly different between TDN and LAO continuers.

Selected coefficients form the first stage of the two-stage model, translated into the relative odds of receiving TDN (versus LAOs), are presented in Table 13.5. Patients starting therapy in January 1983 were 4.5 times more likely to receive TDN than those starting in July 1982. Similarly, patients hospitalized with CHD in period 2 (i.e., around the time of initiation of treatment) were 2.8 times as likely as other patients to receive TDN. In contrast, residents of metropolitan Detroit were only 30% as likely to receive TDN as those residing elsewhere in Michigan. These results underscore the finding that TDN was indeed prescribed to more severely ill patients, and that its geographical diffusion was uneven.

The two-stage estimates of the effect of choice of therapy, presented in Table 13.6., contrast with the single-stage estimates. These results indicate – for all new starts,

Table 13.5. Selected predictors of choice of therapy, from logistic regression (first stage of two-stage model)

Explanatory variable	Relative odds of receiving TDN vs LAO[a]
Month of start of therapy (Jan 1983 vs July 1982)	4.5 (1.8–11.1)
Any hospitalization with CHD in period 2	2.8 (1.4– 5.5)
Use of calcium channel blockers in period 1	1.7 (0.8– 3.6)
Any diagnosis of heart failure	1.4 (0.9– 2.1)
Resident of metropolitan Detroit	0.3 (0.1– 0.8)

[a] Antilogarithm (base e) of the logistic regression coefficient. Confidence intervals (95%) are in parentheses

Table 13.6. Two-stage regression coefficients for choice of therapy variable

Dependent variable (period 3 utilization)	All new starts		Nonswitchers		Continuers	
	Coeff.	(P)	Coeff.	(P)	Coeff.	(P)
Hospital charges	−1.442	(0.611)	−1.342	(0.656)	1.569	(0.555)
Physician charges	−0.455	(0.693)	−0.612	(0.627)	0.037	(0.975)
Total charges	−0.194	(0.825)	0.259	(0.786)	0.263	(0.726)
Hospital days with CHD	0.639	(0.273)	0.110	(0.845)	0.643	(0.156)

nonswitchers only, and continuers only – no significant difference in utilization between TDN- and LAO-treated patients. Furthermore, in moving from single-stage to two-stage estimation, the signs of several coefficients were reversed, from positive to negative. Our best estimates, therefore, suggest that the use of TDN is not associated with different overall levels of health care utilization than the use of LAOs.

Discussion

Data from the Michigan Medicaid program suggest that, in 1982, patients who received TDN were more severely ill than those who were prescribed LAOs. TDN-treated patients were more likely to have had a diagnosis of heart failure, and had significantly higher levels of pretreatment utilization (for example, more frequent and longer hospitalizations and more cardiovascular medications). TDN was also more likely to be prescribed for patients who were hospitalized for treatment of CHD. Not surprisingly, these differences in disase severity were reflected in higher levels of utilization during treatment among TDN patients.

A prestest–postest design was employed to control for such differences between the TDN and LAO patient groups. TDN users had higher levels of utilization during treatment, but their pretreatment utilization was also higher. Pretreatment measures and other potential confounding variables were included, along with choice of therapy, in multiple regression analysis of treatment-period utilization. Our regression results indicate that TDN use is not associated with significantly higher or lower levels of health care utilization or expenditures.

We regard these estimates as only preliminary, however, since the potential for biased results in such an analysis is not obviated by the inclusion of pretreatment indices [3]. There may well have been important differences between patients receiving TDN and those receiving LAOs that were *not* reflected in pretreatment utilization measures. A common source of bias in evaluation studies, termed "maturation," is the process of spontaneous temporal change in some important characteristic of the study subjects. If a maturation process that affects the outcome is more pronounced in one group than the other, then this might be falsely interpreted as an effect of treatment. Another potential problem, termed "local history," is the presence in one group but not the other of some factor or event that affects the outcome, and which is associated with the start of treatment (but is not present during the pretest).

A cause of bias involving both maturation and local history is suggested by the differences noted between TDN- and LAO-treated patients. In our study, all patients began therapy between July 1982 and January 1983, soon after TDN had been introduced to the U.S. market. Many of the physicians who first prescribed TDN were specialists in cardiovascular medicine (unpublished data, Ciba-Geigy Corp.) who, relative to other physicians, may have been caring for sicker patients and may have had more costly practice patterns (e.g., ordered more tests and admitted patients to hospitals more frequently). Thus, if "the sick get sicker quicker" – an example of a maturation process – then utilization levels among TDN-treated patients may have increased more than those among LAO-treated patients for reasons unrelated to choice of therapy. Similarly, if more TDN patients were referred for treatment to specialists in cardiovascular medicine, then this could constitute an important difference in local history, relative to LAO-treated patients. Costly practice patterns of these physicians would *not* have been observed in the pretreatment period, and hence not controlled for in the analysis. Either of these phenomena would be reflected in seemingly higher levels of utilization among patients who received TDN, which could mask any true effect of TDN treatment.

Other limitations of this study may have affected our results. For instance, Medicaid claims data may be less reliable than data collected solely for research purposes. Also, despite a patient population of nearly 800, our power to detect differences in utilization between TDN and LAO patient groups was not great. For example, we would have had less than 80% likelihood of detecting – at conventional levels of statistical significance ($P < 0.05$) – a true 30% difference in utilization levels.

In summary, therefore, our analysis suggests that there are no differences in overall health care utilization between patients receiving TDN and those receiving LAOs. We think it is important to note, however, that TDN need not result in a decline in health care expenditures to be "worth the cost." Physicians or patients may value TDN more highly than alternative antianginal therapies for nonpecuniary reasons. For example, it may be the case that patients prefer the greater convenience of a once-a-day patch, or that physicians believe that compliance improves when patients use TDN. Thus, while we sought in this study to determine whether the use of TDN affects the overall cost of patient management, any future evaluation might also consider other potential benefits of TDN.

Postscript

In the discussion above, we hypothesized that the comparison of treatment-period utilization between patients receiving TDN and those receiving LAOs may have been biased by substantial differences between these patient groups in 1982. This bias, which was likely associated with TDN's introduction to the U.S. market, may not have been adequately controlled by the use of a pretest–posttest design. Since the between-group differences were observed early in the process of diffusion of TDN throughout the medical community, the profile of patients who received TDN in later years may have more closely approximated that of patients who received LAOs. To assess the possibility of a diminishing (or possibly vanishing) "diffusion bias," we compared differences in patient characteristics and levels of health care utilization among cohorts beginning TDN and LAO therapies in 1982 and 1983.

From the 1982–1983 Michigan Medicaid data base, we identified all patients who had started therapy with TDN or LAOs during the periods July-December 1982 (the "1982 cohort") or July-December 1983 (the "1983 cohort"). We defined as "new starts" those patients for whom the first claim for TDN or an LAO was preceded by a 6-month period with no such claim. TDN and LAO patients within each cohort were then compared on a number of variables to assess any change over time in the differences between them.

For each cohort, we examined differences in patient characteristics and in levels of utilization. Patient characteristics included demographic (age, sex, and race) and comorbidity (any diagnoses of hypertension and/or heart failure) variables. Utilization levels were measured during the 6th to 2nd months prior to treatment, and during the month in which treatment began and the prior month (i.e., periods 1 and 2, as defined in the foregoing study). Pretreatment utilization measures included monthly health care charges (total, hospital, outpatient, and drug), occurrence of hospitalization with a diagnosis of CHD, and the use of selected cardiovascular medications (calcium channel blockers, beta-blockers, and nitrates except TDN and LAOs). Start-of-therapy utilization was measured by the occurrence of hospitalization (with any diagnosis and with a diagnosis of CHD).

Table 13.7. Disease severity among TDN vs LAO patients, 1982 vs 1983 cohorts

Variable and cohort	Patient group		Difference	Change in diff. (1982 to 1983)	
	TDN	LAO	(TDN-LAO)	Amount	Signif.
Comorbidity					
Heart failure (% diagnosed):					
1982	54.7	36.5	18.2	− 2.3	NS
1983	41.2	25.3	15.9		
Pretreatment utilization					
Hospitalization with diagnosis of CHD (%):					
1982	12.4	5.6	6.8	− 2.9	NS
1983	9.4	5.5	3.9		
Calcium channel blocker: use (% with Rx)					
1982	13.0	3.1	9.9	− 7.7	P < 0.05
1983	11.4	9.2	2.2		
Nitrate use, except TDN and LAO (% with Rx):					
1982	16.8	14.2	2.6	+ 2.8	NS
1983	17.9	12.5	5.4		
Total drug charges (mean $):					
1982	39	24	15	− 17	P = 0.0001
1983	28	30	− 2		
Total health care charges (mean $):					
1982	851	446	405	−179	NS
1983	727	501	226		
Start-of-therapy utilization					
Hospitalization with diagnosis of CHD (%):					
1982	29.2	9.3	19.9	− 11.9	P < 0.005
1983	22.5	14.5	8.0		
Hospitalization with any diagnosis (%):					
1982	46.0	19.9	26.1	− 14.4	P < 0.01
1983	37.0	25.3	11.1		

To test the hypothesis of diminishing diffusion bias, differences in patient characteristics and levels of utilization between TDN und LAO patients were examined for each cohort. Multiple regression analysis was used to test the statistical significance of any changes in such differences between 1982 and 1983. Each measure was regressed on binary variables representing treatment (TDN versus LAO patients), year (1983 versus 1982 cohorts), and the interaction of these variables (i. e., TDN patients in 1983 versus the remaining patients). The "changes in differences" were measured as the estimated coefficients of the interaction terms; Student's t-test was used to determine their statistical significance.

The 1982 cohort was comprised of 161 TDN and 647 LAO patients; the 1983 cohort included 262 TDN and 392 LAO patients. The cohorts were compared on 15 demographic, diagnostic, and utilization variables. Decreasing dissimilarity was found for 13 of those variables; in four cases the change in differences was significant. Comparisons of the 1982 and 1983 cohorts on selected variables are presented in Table 13.7.

The difference between TDN and LAO patients in the prevalence of heart failure dropped by 2.3 percentage points (from 18.2% to 15.9%). Differences in a number of measures of utilization prior to treatment also decreased. The difference in total monthly health care charges declined by roughly one-half, although this change was not significant. The difference in monthly drug charges diminished significantly, by $17 (from $15 to −$2); the decline of 7.7 percentage points (from 9.9% to 2.2%) in the difference in the use of calcium channel blockers was also significant. Differences in the incidence of hospitalization at the start of therapy (any admission at all, and any for the treatment of CHD) also declined significantly between 1982 and 1983. In 1982, the percentage of TDN-treated patients with any admission at this time was 26.1 points higher than that of LAO patients, but this difference declined by 14.4 points (to 11.7%) in 1983. The difference in the percentage of patients admitted with a diagnosis of CHD declined by 11.9 points (from 19.9% to 8.0%) between 1982 and 1983.

These results support the hypothesis that differences between patients receiving TDN and those receiving LAOs should have diminished over time, as a result of the continued diffusion of TDN throughout the medical community. While differences between TDN and LAO patients remained in 1983, their magnitude often declined significantly from 1982. We therefore conclude that a study of patients receiving TDN of LAOs in 1983 would have a greater likelihood than the foregoing study of yielding unbiased estimates of the true effect of TDN therapy on health care utilization.

References

1. Gordon T, Garst CC (1965) Coronary heart disease in adults, United States, 1960–1962. Vital and health statistics, series 11: data from the health examination survey, no 10. United States. National center for Health Statistics.
2. Rothlesberger FJ, Dickson WJ (1939) Management and the worker: an account of a research program conducted by the Western Electric Company, Hawthorne Works, Chicago. Harvard University Press, Cambridge
3. Cook TD, Campbell DT (1979) Quasi-experimentation: design and analysis issues for field settings. Houghton Mifflin, Boston
4. Barnow BS, Cain GC, Goldberger AS (1983) Issues in the analysis of selectivity bias. In: Stromsdorfer EW, Farkas G (eds) Evaluation studies review annual, vol 5. Sage, Beverly Hills, pp 43–59

14. Cost-Benefit Analysis of Angina Prophylaxis in the Federal Republic of Germany

P. O. Oberender

Subject of Study

The subject of the study was the transdermal therapeutic system, Nitroderm TTS, which has recently been developed for the long-term prophylaxis of angina pectoris attacks. The aim of the study was to determine whether the use of Nitroderm TTS represents a therapeutically and economically useful alternative to the preparations conventionally used for the same indication, namely the long-acting oral nitrates.

Besides the clinical criteria of efficacy and the subjective well-being of the patient, economically relevant effects also served as parameters for the comparison of the two therapies. Parameters considered ranged from the effects on the type, amount, and costs of medication, doctors' services, and medicotechnical services through the frequency, duration, and cost of hospital stays to absences from work due to illness and the resulting loss in economic productivity.

Patients and Methodology

The study was designed as an open, prospective, multicenter-randomized crossover trial to compare Nitroderm TTS with long-acting oral nitrates (LAOs).

Thirty-four privately practising general practitioners and internists took part in the trial. It included a total of 296 patients suffering from chronic stable angina pectoris with coronary heart disease in NYHA stages II and III. Preconditions for admission to the study were a typical case history with at least three stress- or cold-induced attacks per week prior to any previous treatment and S-T segment depression of 2 mm or more that was reproducible in an exercise electrocardiogram. Further criteria for admission to the study were the patient's responsiveness to sublingual glycerol trinitrate or isosorbide dinitrate (ISDN) for the relief of acute attacks and the use of an antianginal combination therapy with at least two different medications, including an LAO.

Patients were excluded on the basis of unstable angina pectoris, pregnancy, valve defects, threatened infarction or actual infarction in the previous 3 months, clinically relevant cardiac arrhythmias, uncompensated heart failure, or severe renal insuffi-

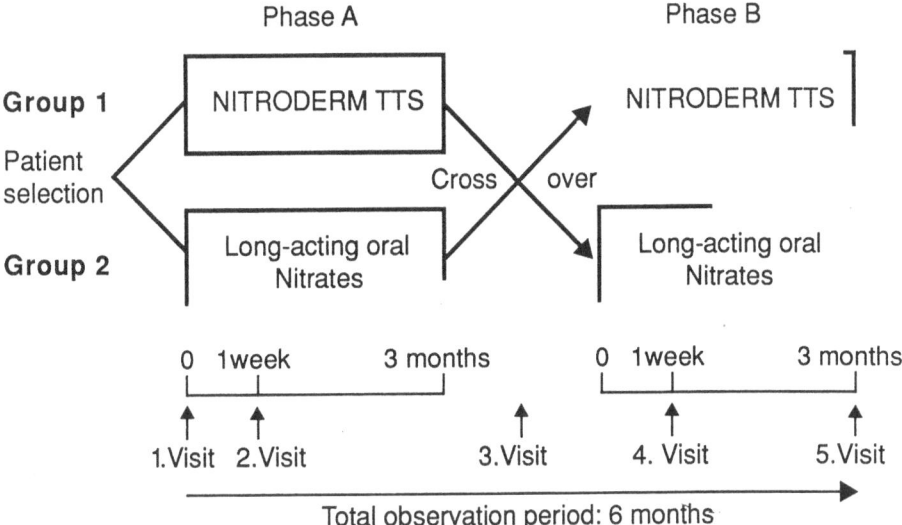

Fig. 14.1. Study design

ciency (creatinine above 2 mg/100 ml). The trial also excluded patients who were treated for their hypertension with beta-blockers or calcium antagonists, uncooperative patients, and patients with conditions that would have impaired compliance (e. g., mental illness, alcoholism). Further exclusion criteria were hypersensitivity to nitro compounds, allergy to patch, anemia, and migraine.

The total duration of the study was 2×3 months, in which the patients, after primary randomization, received either Nitroderm TTS 5 or 10 or LAOs for the prevention of angina pectoris attacks (Fig. 14.1). Any other medication for heart disease was initially continued, but at each scheduled consultation it was considered whether additional coronary therapeutic agents needed to be given or which ones could be discontinued or adjusted in their dosage.

The physicians were instructed to record the following parameters:
- Frequency and severity of angina pectoris attacks
- Tolerability/side-effects of the medication
- Consumption of attack prophylaxis medication (LAOs and Nitroderm TTS)
- Consumption of acute medication
- Consumption of antianginal medication (beta-blockers, calcium antagonists)
- Consumption of non-antianginal medication related to angina pectoris (hypnotics, anxiolytics, tranquilizers, antidepressants, etc.)
- Number of and reasons for spontaneous consultations
- Use of medicotechnical services (ECG, laboratory, etc.)
- Number of days of illness
- Emergencies (home visits, emergency doctor consultations)
- Number of admissions to hospital for angina pectoris and duration of stay

These data were determined at the five consultations scheduled in the trial plan. In order to create a reference basis, the investigating doctors were also asked to determine, from their patient files, selected test parameters for the 3 months *before the start of the trial.*

In addition, at each scheduled consultation the patients filled in a "scale of well-being." This consisted of pairs of opposite attributes (e. g., zestful, lethargic) and its purpose was to evaluate changes in well-being as a result of the medication.

Statistical evaluation of the data was performed with the "Statistical Package for Social Sciences" (SPSS). According to the data type the t-test, the chi-squared test, or the Mann-Whitney method were used.

Results

Comparability of the Treatment Groups

Table 14.1 shows the demographic and clinical data for the two patient groups. No statistically significant differences are apparent which would have been inconsistent with the comparability of the two therapeutic groups.

Table 14.1. Demographic and clinical variables in the two treatment groups

Variable	Long-acting oral nitrates ($n = 146$)	Nitroderm TTS ($n = 150$)
Demographic variables		
Sex (%)		
Males	49.3	51.3
Females	50.7	48.7
Age (years)	64	65
Weight (kg)	72	72
Height (cm)	168	168
Professional status (%)		
Employed	30.1	27.5
Unemployed	4.1	0.7
Not working	65.8	71.8
Health insurance (%)		
AOK	70.6	62.7
Ersatzkasse	22.6	29.3
Private insurance	6.9	8.0
Clinical variables		
Severity of CHD (%)		
NYHA II	64.4	62.0
NYHA III	35.6	38.0
Duration (months)	54.2	57.4
Occurrence of angina attacks		
Only at day time	49.3	51.4
Only at night	2.1	2.0
Day and night	48.6	46.6
Blood pressure (mmHg)		
Systolic	143.8	141.0
Diastolic	85.2	83.6

Symptoms of Angina Pectoris

In phase A, markedly fewer angina pectoris attacks were recorded with Nitroderm TTS than with the LAOs (Fig. 14.2). By comparison with the previous medication, in the corresponding patient group I the average frequency of attacks was already reduced from 4.8 per patient per week to 3.2 after 1 week and further to only 1.9 attacks after 3 months. In the same trial phase the frequency of attacks in group II on LAOs also fell, but only from an average of 4.5 attacks to 4.0 and 3.0 respectively.

After the changeover of medication – i.e., in *phase B* – the frequency of attacks in patient group I, now on LAOs, deteriorated slightly from 1.9 to 2.3 and remained steady at this level for the duration of the trial. In patient group II, on the other hand, on Nitroderm TTS there was a further marked reduction in the frequency of attacks.

On the adhesive patch therapy the patients not only had fewer attacks but these were also of shorter duration and milder. Differences were already apparent in this connection after 1 week of treatment.

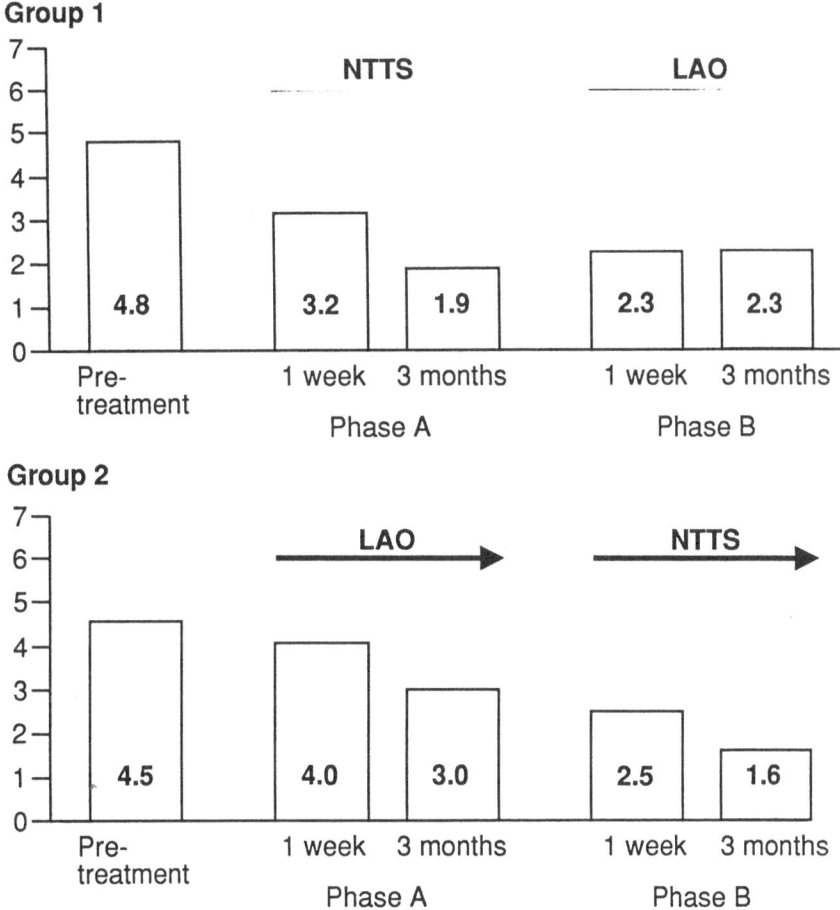

Fig. 14.2. Changes in the average number of attacks, per week and per patient. *LAO,* long-acting oral nitrates; *NTTS,* Nitroderm TTS

The difference was particularly striking in those patients who had complete cessation of attacks: at the end of the 3-month period of treatment, 34% of those initially treated with Nitroderm TTS were free from attacks compared with only 13% in the reference group. After crossover the percentage of attack-free patients increased from 13% to 38% in those receiving Nitroderm TTS in phase B while in those receiving LAOs it fell from 34% to 30%.

As regards the number and severity of side-effects, there were no substantial differences between the two alternative therapies although side-effects differed in their nature: Nitroderm TTS elicited skin reactions in roughly 10% of the cases, and the LAOs in just under 3%. (Nitrate) headaches occurred in 10% of the Nitroderm TTS patients, compared with 18% of patients on the LAOs. In both groups most of the side-effects were mild and transient.

Change in Well-being

The efficacy and tolerance findings are reflected in the subjective well-being of the patients. The test results for the scale of well-being used in the trial are shown in Fig. 14.3. A lower score corresponds to an improvement in well-being.

In *phase A* both patient groups started with almost the same degree of impairment. On Nitroderm TTS, during this phase the average overall score was reduced from 53.7 points to 36.9 points (i. e., by 16.8 points), but on LAOs only from 52.9 points to 46.8 points (i. e., by 6.1 points). Thus the reduction in the impairment on Nitroderm TTS in phase A was more than two and a half times as great as on LAOs.

In *phase B* on Nitroderm TTS the average overall score was further reduced from 46.8 points to 34.3 points (i.e., by 12.5 points), while on LAOs the average score already rose in the first week of treatment from 36.9 points to 42.5 points (i. e., by 5.6 points), remaining at this increased level until the end of the trial.

The absolute difference in the change in the average overall score in phase B was thus 18.1 points in favor of Nitroderm TTS (all differences are highly significant with $P < 0.01$).

Costs of Medication

Within the framework of the trial, the total drugs prescribed were divided into four groups:
- Drugs for the prevention of angina pectoris attacks (Nitroderm TTS or LAOs)
- Drugs for the treatment of acute angina pectoris attacks (nitroglycerin)
- Drugs for the treatment of the underlying coronary disease (beta-blockers, calcium antagonists)
- Other drugs for the treatment of symptoms (tranquilizers, analgesics, etc.)

Analysis of the costs of medication in the two therapy groups over the whole of the trial period (2×3 months) shows the expected pattern in the area of attack prophylaxis (Table 14.2). Per patient, treatment with Nitroderm TTS over the 6-month period was more than twice as expensive as treatment with LAOs (DM 523.58 vs DM 241.51). This was due to the comparatively high daily dose costs of the preparation.

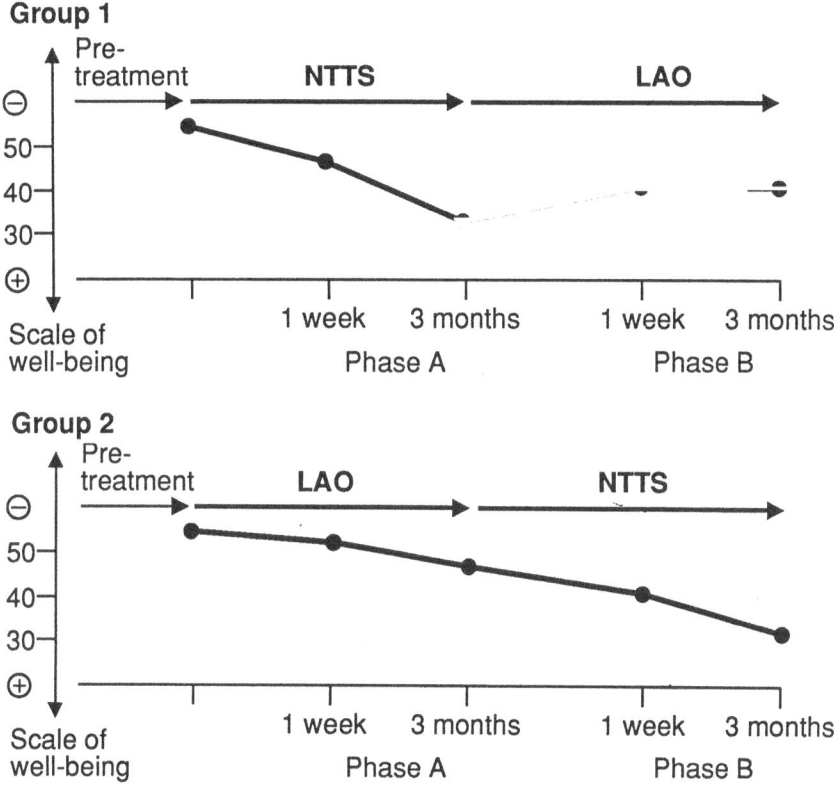

Fig. 14.3. Changes in the subjective well-being of patients. *LAO,* long-acting oral nitrates; *NTTS,* Nitroderm TTS

Table 14.2. Costs of Medication (in DM) for 180 days of treatment per patient

Type of medication	Nitroderm TTS	LAO	Difference, absolute	Difference as % of LAO
Angina prophylaxis	523.58	241.51	+ 282.07	+ 116.8
Treatment of acute attacks	8.28	11.92	− 3.64	− 30.5
Antianginal medication	213.39	226.65	− 13.26	− 5.9
Symptomatic medication	17.99	22.35	− 4.36	− 19.5
Total	763.24	502.43	+ 260.81	+ 51.9

On the other hand, with regard to medication for the treatment of acute attacks, Nitroderm TTS was found to have a cost advantage (DM 8.28 vs DM 11.92). The reason for this was the highly significant reduction in the number of attacks on Nitroderm TTS compared with the reference therapies.

The better efficacy of Nitroderm TTS evidently also induced the physicians treating the patients to reduce the number and/or dosage of other antianginal medications in the form of beta-blockers and calcium antagonists, and also of non-antianginal adjuvant medication such as tranquilizers, hypnotics, and analgesics. The costs of antianginal medication were therefore roughly 6% lower on Nitroderm TTS than on LAOs (DM 213.39 vs DM 226.65) and the costs of symptomatic medication roughly 20% lower (DM 17.99 vs DM 22.35).

These saving effects reduced but did not cancel out the additional cost of Nitroderm TTS in the area of attack prophylaxis. All the same, if all expenditure on medication is taken into account the extra costs for attack prophylaxis were reduced from 117% to 52%.

Doctors' Services

Table 14.3 compares the costs of doctors' services in connection with the treatment of angina pectoris in the two test groups over the 6-month period of the trial: consultations, thorough examinations, ECGs, injections, etc. The official tariff schedules for doctors (GOÄ and BMÄ) served as the basis for the calculation.

In view of the better clinical efficacy of Nitroderm TTS, the attending physicians evidently found it possible with this treatment to reduce the number of medical checks, home visits, and medicotechnical services provided.

For consultations, in the 6-month trial period on LAOs the average cost per patient was DM 87.24, compared with DM 68.23 on Nitroderm TTS. The cost of home visits per patient on LAOs was DM 44.07 compared with only DM 25.21 on Nitroderm TTS. For emergency visits, the cost in the Nitroderm TTS group was slightly higher than in the LAO group (DM 5.48 vs DM 4.93). In the area of medicotechnical services, on the other hand, the cost difference in favor of Nitroderm TTS was almost DM 39.00.

Overall, the costs of doctors' services for the patients treated with Nitroderm TTS over the whole trial period amounted to DM 183.74, roughly 30% less than for the patients on LAOs (DM 259.87).

Stays in Hospital

Among the 296 patients, five cases of hospitalization were reported in the study period. In the Nitroderm TTS group three cases were recorded in the first 3-month period but none in the second 3-month period, while in the group on LAOs only one case was recorded in each period. In the three patients on Nitroderm TTS, the stays in

Table 14.3. Costs of doctors' services (in DM) for 180 days of treatment per patient

Type of service	Nitroderm TTS	LAO	Difference, absolute	Difference as % of LAO
Regular consultations	68.23	87.24	− 19.01	− 21.8
Home visits	25.21	44.07	− 18.86	− 42.8
Emergency treatment	5.48	4.93	+ 0.55	+ 11.2
Medicotechnical services	84.82	123.63	− 38.81	− 31.4
Total	183.74	259.87	− 76.13	− 29.3

hospital were of 13, 16, and 18 days; in the two patients on LAOs, hospitalization was for 10 and 20 days. Thus a total of 47 days were spent in hospital by Nitroderm TTS patients as compared with 30 by patients on LAOs.

In order to determine the cost of the days in hospital, the study used the average daily charge of all hospitals in the Federal Republic of Germany, namely DM 260.60 (1985). On this basis, the hospital costs with Nitroderm TTS were found to total DM 12248.– and those with LAOs DM 7817.70, or, respectively, DM 81.65 and DM 52.60 per patient.   Thus in this area, the costs with Nitroderm TTS are more than one and a half times as great as with the reference medication.

Days of Absence from Work

Eighty-five of the 296 patients included in the trial were engaged in gainful activity (29%). Although the two therapeutic groups included almost equal numbers of gainfully employed persons, there were substantial differences between the therapeutic groups in the number of working days lost. On Nitroderm TTS, there were only 12 cases with absences as compared with 41 periods of work loss by patients on LAOs (see Table 14.4).

The difference in the total days of absence from work is correspondingly large. Over the whole trial period the Nitroderm TTS patients stayed away from their jobs for a total of only 52 days, while those treated with LAOs remained away from work for 297 days – well over three and a half times as long.

Table 14.4. Loss of working days and productivity

Phase	Nitroderm TTS			LAO		
	A	B	A + B	A	B	A + B
Number of patients	150	146	296	146	150	296
Absences from work *with* a doctor's certificate						
Number of cases	3	3	6	11	10	21
Duration in days	20	23	43	130	75	205
Absences from work *without* a doctor's certificate						
Number of cases	4	2	6	9	11	20
Duration in days	6	3	9	30	62	92
Absences from work (total)						
Number of cases	7	5	12	20	21	41
Duration in days	26	26	52	160	137	297
Loss in productivity (DM)						
Total	5149	6788	11937	29740	25026	54766
Per patient	34.3	46.5	80.8	203.7	166.8	370.5

Table 14.5. Total costs of therapy (in DM) for 180 days of treatment per patient

Type of costs	Nitroderm TTS	LAO	Difference, absolute	Difference as % of LAO
Medication	763.24	502.43	+ 260.81	+ 51.9
Medicotechnical services	183.74	259.87	− 76.13	− 29.3
Hospitalization	81.65	52.60	+ 29.05	+ 55.2
Loss in productivity	80.81	370.54	− 289.73	− 78.2
Total	1109.44	1185.44	− 76.00	− 6.4

For the evaluation of this absence from work and in order to establish the loss of productivity for the economy, the average income for the occupations exercised by the relevant patients was taken as a basis. The occupation was stated on the personal history sheet; the average income for the occupational category was obtained from a survey of 4000 people carried out in the Federal Republic of Germany by Infratest/ KFM in 1986. On the basis of this evaluation, the difference in favor of treatment with Nitroderm TTS in terms of costs was even greater than in terms of days. For patients on Nitroderm TTS, the loss of productivity over the whole test period was valued at about DM 12000, and for those on LAOs at about DM 55000, i. e., more than four and a half times as much.

As a result of treatment with Nitroderm TTS, patients under study could produce or supply more goods and services, to a value of roughly DM 43000. This, however, assumes that they were not unemployed.

Total Costs of Therapy

Thus, if we consider the overall costs of treating angina pectoris with Nitroderm TTS, we see a relative increase in the costs of medication and hospital stays. However, Nitroderm TTS therapy more than compensates for this increase by savings in the fields of doctors' services and lost productivity. As a result, it ultimately reduced the total costs of therapy by more than 6% (see Table 14.5).

15. Discussion of Part II

The presentations in Part II of the symposium summarized the results of four studies on the economic consequences of drug therapy, taking Nitroderm TTS as an example. In an invited discussion, the author (*B. R. Luce*) criticized the four studies from Austria, Germany, the United States of America (Michigan) and Switzerland on the economic consequences of Nitroderm TTS therapy:[1]

It is fortunate that there are multiple economic studies which report on the same topic – a rarity among this type of research. Although the four studies provide information about the economic consequences of one drug, the research designs are different, populations are different, the time periods differ somewhat, and the clinical settings differ. However, the same fundamental research question was asked and the same basic comparisons were made. Together, the studies help us interpret the generalizability – or, in methodological terms, the external validity – of any one study by itself.

As a rule of thumb, when evaluating medical interventions, the more rigorous the research design, the less the observed effect. All pharmaceutical firms know this to be true: the early promise of initial, uncontrolled clinical studies nearly always has to be discounted to some degree. Thus, phase 2 randomized, controlled trials required by the Federal Drug Administration (FDA) of the United States commonly result in less dramatic claims of efficacy. Since we have four studies of varying designs of rigor, the phenomenon of more rigorous research design leading to less observed effect can be tested.

The Framework of the Analysis

The papers of Part II will be discussed in order of the degree of rigor of the research designs. Table 15.1 presents them in that order with the German study first, the Swiss last.

Both the Michigan (Oster et al.) and the Swiss studies use data from 1982 and 1983, while the German (Oberender) and Austrian (Dinkel) studies use 1985 or 1986 data. The three European studies examined data taken over a 6-month period; the US study examined data recorded over 18 months. All studies can be considered to have taken the perspective of the insurer; that is, they all considered total health care utilization

---

[1] Nitroderm TTS is also referred to as Transderm Nitro (TDN) in this discussion

Table 15.1. Framework for the analysis

	German	Michigan Medicaid	Austrian	Swiss
Year of data	1986	1982–1983	1985–1986	1982–1983
Duration of study (months)	6	18	6	6
Comparison	← Long-acting oral vs. Transderm Nitro →			
Sample size				
LAO	146	622	–	–
TDN	150	146	63	25
			(Pre-post)	(Pre-post)
Perspective	← Insurance →			

LAO = Long-active oral
TDN = Transderm Nitro (or, in Europe, Nitroderm TTS)

and the respective costs. All studies compared long-acting oral nitrates (LAO) to transdermal nitroglycerin (TDN) therapy, and all studies were conducted by independent research firms. Table 15.1 presents comparative information.

Study Design

The *German study* was designed as an open, prospective, two-arm multicenter, randomized crossover trial comparing long-acting orally administered (LAO) nitrates with the transdermal nitroglycerin patch (TDN). The 296 patients were recruited by 34 private practitioners. The patients were randomly assigned to either the LAO "control" group or the TDN "experimental" group. After 3 months, the patients in the LAO group were switched to TDN, and vice versa. All health care utilization was monitored, including pharmaceuticals, visits to or from a physician, ancillary services, and hospitalizations. In addition, patients filled out quality-of-life questionnaires and reported the number of days of illness requiring absence from work.

The *Michigan Medicaid Study* is a two-arm, retrospective analysis of 768 patients who had been prescribed either LAO or TDN. To be included in the study, patients had to have had a complete 6-month health care utilization history prior to receiving either LAO or TDN. Their utilization patterns and resulting charges were then followed (retrospectively) for 12 months. All medical care utilization during the treatment and pretreatment period was analyzed.[2]

A number of statistical techniques were used to control for selection bias. First, in order to avoid a bias due to patients' being assigned to a group because of drug failure, the sample included only patients who had no recent history of being prescribed either LAΘ or TDN. That is, only "new starts" were included. Further, demographic differences such as age, sex, race, and place of residence were controlled. To control

---

[2] The analysis specifically excluded the 30 days prior to and the 30 days after the point at which treatment was initiated because the authors believe that the initiation of a therapy is often characterized by hospitalization, which could bias the analysis.

Table 15.2. Study design

	German	Michigan Medicaid	Austrian	Swiss
Research design	Cost-effectiveness Prospective Two-arm Crossover	Cost comparison Retrospective Two-arm New-start	Cost Retrospective One-arm Crossover	Cost comparison Retrospective One-arm Crossover
Control	Randomized	Statistical, historical	Historical	Historical
Data collection	Physicians recorded	Computerized claims files	Physicians interviews	Physicians interviews

for severity of illness, a number of measures were used including monitoring prior utilization histories and comorbidities. Prior utilization control was particularly important, since the TDN group had a much higher history of medical care utilization, and patients in this group were therefore probably much more ill. It was not possible to control for physician specialty.

The authors used both an ordinary least square regression model, plus a "two-stage" regression technique, the latter to control better for selection bias.

The *Austrian study* by Dinkel was an open, retrospective, multicenter crossover design consisting of 63 men and women who had been switched by their physicians from oral medication to TDN. At the time of the study, TDN had just been introduced into the country. Utilization and cost data were obtained using a structured physician questionnaire during an oral interview.

The *Swiss study,* using 1982–1983 data, was smaller ($n = 25$), but the design was identical with that of the Austrian study. The medical utilization history of patients who had been switched by their physician from LAO to TDN was analyzed before and after the change in medication. The study was retrospective and data was collected via structured physician interviews.

Results

Although there were mixed results from the four studies, they were not necessarily inconsistent with one another. The general hypothesis across all studies had been that the increased costs of prophylactically treating angina using the nitroglycerin patch rather than LAO would be partially or wholly offset by decreased costs of medical services in the nitroglycerin patch group. All studies found an increase of cost in the TDN group due to anti-angina prophylaxis (an expected result since the nitroglycerin patch is more expensive than LAO medication). Two of the four studies (Austrian and Swiss) indicated net savings due to lower medical care utilization in the patch group. The Michigan study showed no significant difference in medical care costs for the two groups, and the German study indicated slight savings in the TDN group if one also considers indirect costs (i. e., earnings at work lost or gained). Specific comments for each study follow.

German Study

The net cost differential indicated a small total cost advantage of using the TDN patch compared to the long-acting oral medication. This net saving was due, in large part, to savings in work loss. Net medical costs were significantly higher for the TDN group. However, a closer look at utilization patterns reveals evidence supporting the study hypothesis. For instance, as expected, prophylaxis was more expensive per patient for the TDN group (DM 524 vs. DM 242), but these costs were somewhat offset by a 9% decrease in other pharmaceutical costs (DM 231 vs. DM 249), a 27% reduction in physicians' visits (DM 99 vs. DM 135) and a 31% reduction (DM 85 vs. DM 124) in ancillary "medico-technical" services. As counter to these savings, the TDN group had three hospitalizations compared to two hospitalizations for the LAO group, thus severely affecting the comparison of medical care costs (DM 13195 for TDN; DM 8579 for LAO).

However, significant savings were estimated from productivity gains due to less angina-related work loss in the TDN group, who were off work for only 52 days, compared with 197 days in the LAO group. The authors calculated the resulting savings by estimating the expected income loss to each individual participant in this study. Results indicated that using this method, the 52:197 (1:3.7) days work-loss ratio translated to a DM 12000:DM 55000 (1:49) earnings ratio. The differential between the work-loss ratio and the earnings ratio, although not explained in the manuscript, may be due to the TDN group being in a higher income bracket.

When the authors combined net direct and indirect costs, results indicated a net saving of approximately six percent for the TDN group. No statistical tests of significance were reported.

The Michigan Study

This study clearly indicated that TDN was prescribed for more severely ill patients than was LAO. Thus, the two groups whose subsequent utilization was compared were different in a very fundamental way.

Having controlled for as many variables as possible, including the use of a two-stage regression technique, the authors conclude that TDN use leads to neither higher nor lower overall health care costs.

The initial univariate analysis indicated that TDN patients had higher utilization before, during, and after treatment. While drug costs of the TDN group increased during the treatment period, as expected, total costs remain fairly constant, indicating that non-drug costs declined slightly, while LAO showed a definite increase in utilization. Thus, the fundamental hypothesis seemed to be supported. However, increasingly sophisticated analytical techniques indicated that TDN use may be associated with slightly higher utilization. However, results were not on the whole statistically significant.

The Austrian Study

The results of this study indicated that the 88% increase in costs of angina prophylaxis with TDN was just slightly more than offset by a decrease in medical care costs due to
a) drug therapy for the treatment of angina,
b) adjuvant drug therapy,
c) physicians' visits, and
d) ancillary services.
The authors report that the TDN group experienced a 4.1% decrease in total costs compared to the LAO group. No statistical significance tests were reported.

The Swiss Study

This smaller study reported similar but greater cost savings with TDN. For instance, the increase in costs of prophylaxis for the TDN group was only 11%, but there was an 80% reported decrease in angina attacks in the TDN group compared to the LAO group leading to savings in other pharmaceutical costs (80% for angina therapy, 75% for adjuvant drug therapy, and significantly lower costs due to physicians' visits). Overall, the TDN group experienced a 31% decrease in reported cost. No statistical significance tests were reported.

Discussion

The overall impression one has from reviewing the four studies described in this paper is that the increased cost of TDN prophylaxis compared to LAO is probably at least offset by savings due to the efficacy of the therapy.

The most rigorously controlled study (the German) shows a fairly large increase in medical costs in the TDN group, but this is due mostly to three (TDN) versus two (LAO) hospitalizations, indicating that the number of subjects in the study was probably too small to include hospitalization as an end-point. All other health care utilization including drug therapy, physician visits, and ancillary services showed consistent savings in the TDN group, collectively indicating possible higher efficacy in the TDN group. But even these nonhospital "savings" did not completely offset the higher costs of prophylaxis in the TDN group.

There was a surprisingly large saving due to a decrease in work loss in the TDN group. It is difficult to evaluate those savings in that only 28% of the sample was reported to be productively employed. Assuming the authors only included that subgroup in the work-loss analysis, it may be that by chance the relatively small numbers of LAO workers were the ones within the larger sample who developed complications. A better way to analyze the findings might have been to record the ability to carry out one's normal daily routine activities, whether housework, golf, or paid labor, and to impute wages or value of one's time to each.

Another related concern in interpreting the results is that the investigators report that they valued the individual's income based on the average income for the occupations of the relevant patients. Since data presented in the study indicated that the wages for the LAO group may have been higher than those for the TDN group, it would be important to know what that differential is.

Finally, no statistical analysis was presented, so it is difficult to pass judgment on the significance of the findings.

The Michigan Medicaid study is an excellent example of how to compare two groups statistically when they differ fundamentally from the start. The main problem was that the group that was started on TDN was clearly much sicker than the LAO group. The TDN group had much higher prestudy utilization, including more frequent and longer hospitalizations, were on more intensive cardiac drug therapy, and were more likely to be suffering from heart failure. A nagging concern which the investigators were unable to explore was the possibility that physician practice patterns may have differed between the two groups. Since TDN was new in 1982–1983, it is possible that only cardiologists, or the more aggressive physicians, used the "new" drug. If that were true, then some of the observed effect may be physician-induced bias. However, the Michigan Medicaid database does not identify physician type.

Thus, the Michigan study provides us with some reason to believe that there is neither positive nor negative medical utilization and/or cost effect from using TDN compared to LAO.

Both the Austrian and Swiss studies are less equivocal in their respective findings, but are still difficult to interpret fully. As a one-arm, retrospective crossover study, it offers the opportunity to compare prior and past utilization patterns for the same patient population group.

The concern in interpreting the findings is that one cannot be sure of the reason for the switch in medications. One obvious reason could be that the patient was not doing well on the LAO regimen, which would give rise to obvious bias. The patient could be getting sicker, medication compliance could be poor, or the patient might not be tolerating the LAO drug very well. All these biases are not necessarily a problem if they are known. Conclusions could be limited to these subgroups of patients, because one could not legitimately generalize to others who were doing well on the LAO regimen.

Another reason was an "expectation" on the part of the physician that the patient would do better on TDN. This was reported by physicians in the Swiss study as the reason they switched medications. A related point to be considered is the manner in which data was collected – by interviewing physicians. Physicians who had chosen a new therapy for their patients might be predisposed to report that those patients did better on the new regimen.

Finally, sample sizes were relatively small. The Austrian study was the larger and showed a moderately positive "TDN effect"; the Swiss was a very small study ($n = 25$) and showed the largest positive "TDN effect" of all four studies.

Conclusions

In the final analysis, the four studies examined the economic effect of prescribing the Transderm Nitro (TDN) patch compared to the previously conventional long-acting oral (LAO) nitroglycerin therapy. The most rigorous study (German) indicated that higher utilization costs due strictly to higher costs of angina prophylaxis and one extra hospitalization were partially offset by lower drug, ancillary, and physician utilization, and were more than offset by the reported value of productivity gains due to less work loss. Concerns were noted about the accounting for productivity gains and the statistical significance.

The second most rigorous study (Michigan Medicaid) was not able to show any statistically significant effect on utilization and cost from using TDN compared to LAO. Concern was expressed by the authors about the extent to which it was possible to control for all bias, since the TDN group had much higher pretreatment utilization patterns.

The Austrian study shows a mildly positive utilization "TDN effect" (i.e., lower costs), and the Swiss study showed a highly positive utilization cost decrease from using TDN. Concerns were noted about the small sample size, especially in the Swiss study, the bias due to lack of an adequate control group, and the lack of reported statistical significance tests.

Overall, the hypothesis that the use of Transderm Nitro therapy leads to lower overall health care costs cannot be rejected. All four studies indicated that the increased costs of prophylaxis were offset (to varying degrees) by decreased costs of acute angina-related disease.

In the general discussion that followed the analysis of the four studies presented by Luce from the Battelle Human Affairs Research Centers in Washington, DC, one discussant stressed the importance of the random assignment of patients to the study group (in this case Nitroderm TTS) and to the control group (e.g. LAO). In the Michigan study, for instance, during pretreatment the total charges in 1982 were US $ 851 for the TDN patients vs. US $ 446 for the LAO patients. This was before they ever saw the drug. It is possible, and the example illustrates the point, that the patients in the experimental group were sicker on average than those who were prescribed the traditional treatment.

Another aspect of importance is the relative concept of cost. All studies presented in Part II can be considered to have taken the perspective of the insurer. In many countries health insurances are interested only in direct costs (medical treatment, hospital care, and drug costs) but ignore indirect costs (loss of production and opportunity costs). If indirect costs and gains are important and if there are large and clear differences between direct and indirect costs and benefits, the comparison of both becomes important. Thus, higher direct costs (due to the application of a more expensive drug) may yield a proportionally higher indirect gain which goes undetected if indirect costs are not included in the equation. Insurance companies are, in general, more interested in their own limited perspective than in the broader context of national economy. In a comprehensive socioeconomic evaluation of drug therapy the aspect of indirect costs and benefits quite often becomes of decisive importance.

Part III

Measuring the Effectiveness of Drug Therapy –
Social and Qualitative Criteria

16. Quality of Life – Principles and Methodology

S. R. WALKER

Introduction

At a recent symposium in London, Patrick and Erickson [7] stated that "health care professionals from many specialities and disciplines are becoming increasingly aware that one of the major goals of medical care and technology is to improve patients' quality of life. Enhancing quality of life is as important as other goals of health and medical care such as preventing disease, effecting a cure, alleviating symptoms or pain, averting complications, providing humane care and prolonging life." Historically, the clinical evaluation of new medicines has progressed substantially over the past 40 years. It was in the 1940s and 1950s that the importance of evaluating the effectiveness of medicines was appreciated and the principles and methodology of the controlled clinical trial were established. Teeling Smith [10] elaborated the historical background to the development of economic measurements of health and the effectiveness of therapy by demonstrating (Table 16.1) that this clinical trial methodology answered the question "Does the treatment work?" while cost-benefit analysis, developed in the 1950s and 1960s, provided the framework for measuring the overall economic benefits of medical care and setting them against the corresponding costs by answering the question "Does the treatment pay off?". In the 1960s and 1970s the economic emphasis swung away from cost-benefit analysis to the concept of cost-effectiveness analysis, which answered the question "Which is the most effective

Table 16.1. Development of economic measurements of health and the effectiveness of therapy (Teeling-Smith [10])

Year	Evaluation	The question it answers
1940s/50s	Clinical trials	Does the treatment work?
1950s/60s	Cost-benefit analysis	Does the treatment pay off?
1960s/70s	Cost-effectiveness analysis	Which is the most effective treatment using given resources?
1970s/80s	Cost-utility analysis	How does treatment affect length and quality of life?

treatment using given resources?". However, over the past 10–15 years, the principles of cost-utility analysis have introduced the concept of utility as a measurement of value to replace the financial measures and answer the question "How does the treatment affect the length and quality of life?".

In 1981 Ware et al. [12] stated that there were a number of reasons for studying health status, namely to:

1. Measure the effectiveness of medical intervention
2. Improve clinical decisions
3. Assess the quality of care
4. Estimate population needs
5. Understand the causes and consequences of differences in health

With chronic disabling diseases becoming increasingly evident as other life-threatening diseases have been more or less eradicated in the developed world, the importance of measuring health status in those diseases which have an impact on a patient's quality of life is underlined, such as respiratory diseases, diseases affecting the central nervous system (e. g. Parkinson's disease) and the major cardiovascular problems, including angina. If we are to improve clinical decision-making then we need to have some means of evaluating health-related quality of life outcomes. This is because we need not only to evaluate the impact of diseases on specific populations but also to determine the effectiveness of treatments, including measures of quality of life, which will then allow us to examine resource allocation. Logically, in coming to terms with measuring health status (Fig. 16.1), the first step is to identify some

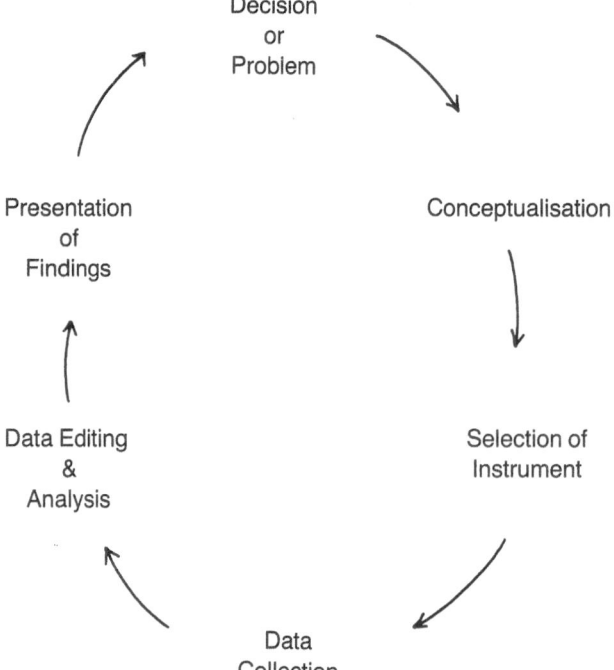

Fig. 16.1. Health status information system (Patrick and Erickson [7])

problem to be solved or decision to be made. A key part of this first step is an explicit recognition of the need for data on health-related quality of life along with the identification of the target population. Conceptualisation involves the identification of all the relevant and possible outcomes of disease and treatment that may be involved in deciding among alternatives. Instruments can then be selected using available methodological and practical criteria and the data subsequently collected. These data can then be edited and analysed using a variety of scoring and statistical techniques, and the final step is to present the findings to decision-makers [7].

Conceptualisation of Health Status

In conceptualising quality of life it has been suggested [9] that there are three major components: physical function, social function and emotional function (Fig. 16.2). Many of the instruments that have been developed in the past 10–15 years allow the components of quality of life to be grouped in this way. Some of the above items can obviously be included in more than one category and cannot be placed under only one heading except as a basic classification. By no means do all health status indices include all these areas, but it is considered that the majority need to be covered for a health-related quality of life instrument to be useful in the context of clinical evaluation.

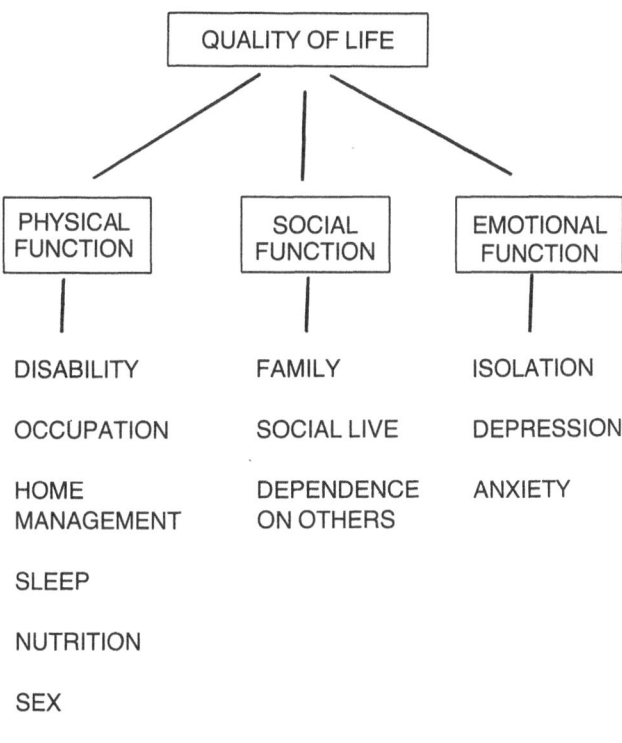

Fig. 16.2. Components of quality of life

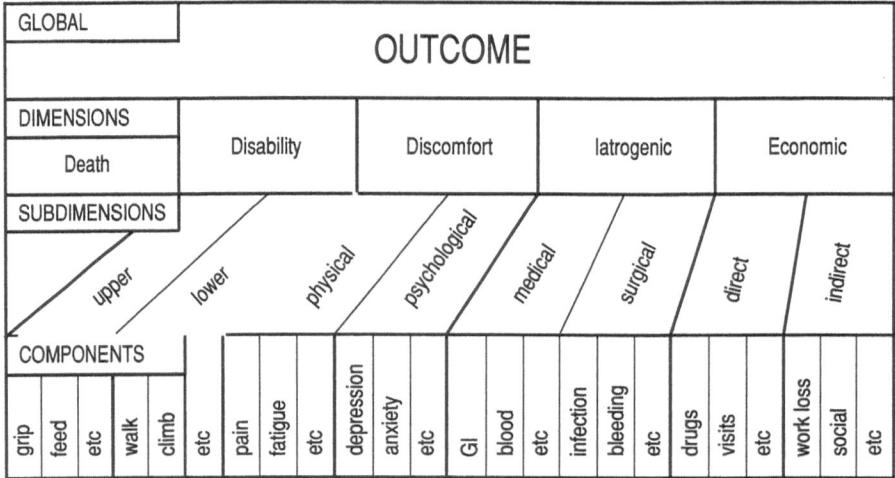

Fig. 16.3. Conceptualisation of "outcome" as a hierarchy of dimensions, subdimensions and components (Fries [3])

Measurement of health status requires the specification of dimensions and items which are to be included in a health status index or profile. Fries [3] provides one approach and suggests a framework for rheumatoid arthritis whereby outcome is conceptualised as a hierarchy of dimensions, subdimensions and components; this is shown in Fig. 16.3.

In the treatment of many diseases, today's goal is not cure but an improvement in function, a resolution of symptoms endeavouring to limit the progression of disease and improvement of the patient's quality of life. Patrick and Erickson [7] state that the concepts of health-related quality of life can be considered on the continuum of well-being scale (Table 16.2) covering such concepts as satisfaction, general health percep-tions, social well-being, psychological well-being, physical well-being and role limita-tions, disease and death. While it is possible simplistically to imagine a visual analogue global scale (Fig. 16.4) where 0 equals dead and 1 equals a state of complete health and well-being (morbidity and mortality improvement being combined into a single weighted scale), this is probably not satisfactory for adequately describing and measuring the relevant components of quality of life.

Fig. 16.4. Morbidity and mortality improvement combined into a single weighted scale

Table 16.2. Concepts of health-related quality of life on the continuum of well-being (Patrick and Erickson [7])

Concept/category	Definition/indicator
Satisfaction	With physical, psychological, social and spiritual well-being
General health perceptions	Self-rating of health, perceptions of health, or worries about health
Social Well-being	
Social integration	Participating in the community
Social contact	Interaction with family and friends
Intimacy	Perceived feelings of closeness/support
Opportunity	Equality of opportunity because of health
Psychological well-being	
Affective well-being	Psychological attitudes and behaviours, including distress, and general well-being or happiness
Cognitive well-being	Alertness: disorientation, problems in reasoning
Physical well-being and role limitation	
Activity restrictions	Acute or chronic limitation in physical activity, mobility or self-care
Limitations in usual roles	Acute or chronic limitations in social roles, school, work, household management, recreation
Fitness	Performance of physical activity with vigour and without excessive fatigue
Disease	
Subjective complaints	Reports of physical and psychological symptoms, sensations, pain, health problems or feelings not directly observable
Signs	Physical examination: observable evidence of defect or abnormality
Self-reported disease	Patient listing of medical conditions or impairments
Physiological measures	Records and clinical interpretation
Tissue alterations	Pathological evidence
Diagnoses	Clinical judgements after "all the evidence"
Death	Mortality; survival

Methodology

There are a large number of instruments available which purport to relate to quality of life. These can be classified as follows:

1. *Population survey tools,* i.e. mortality and morbidity data

2. *Instruments for assessing diagnosis, progress and outcome*
   a) Disease-specific instruments for cancer, rheumatoid arthritis etc.
   b) Population-specific instruments for geriatric groups, rehabilitation pro-
      grammes, etc.
3. *General Measures of health status*
   a) Indices, e.g. Quality of Well-being Scale, Rosser and Watts Index
   b) Profiles, e.g. Sickness Impact Profile, Nottingham Health Profile

Over the years a number of investigators working together in multidisciplinary
teams have developed composite measures of health status. While single measures of
health-related quality of life cover a single health condition, i.e. life or death, the
composite measures summarise two or more aspects of health status or quality of life
in an aggregated index or profile. Global health status indices are expressed as a single
aggregated score whereas health profiles refer to measures in which each component
is assigned a summary score. In certain situations health profiles may also provide an
overall score that summarises across all components (such as the Sickness Impact
Profile). While single indicators of mortality and morbidity may be valuable for
population monitoring and are sensitive to differences amongst population groups,
they are not sufficiently sensitive indicators of population health and well-being to
allow measures of the effectiveness of treatment of diseases.

Table 16.3 itemises four global indices, their features or attributes and the way in
which the data can be aggregated into a single score. In contrast, Table 16.4 gives
details of five health profiles indicating the number and type of dimensions measured
and the total number of items or questions included in each of these profiles.

Composite measures covering more detailed health states and aspects of well-being
may be necessary to detect health changes and classify individuals. No single figure
index can take into account all the components of well-being, satisfaction and opportunity that make up health-related quality of life and they may therefore be inade-

Table 16.3. Global indices

Index name	Features	Index "score"
Bush 43 function levels 36 symptom problem complexes	Four attributes: mobility, physical function, social function, symptom-problem complex	Quality of Well-Being (QWB) for individual: "well-years" can be aggregated
Rosser 29 states	Two dimensions: distress, disability	Utility scales/states for indivuals; can be aggregated
Wolfson 54 items	Ten functional categories: dressing, bathing, continence, eating, under-standing, wheelchair, ambulation, transfer, speech, mental status	Weighted index for individuals, aggregated for group
Torrance 23 items	Four attributes: physical function, role function, social-emotional function, health problem	Individual utility function aggregated to group utility function

Table 16.4. Health profiles

Index	Dimensions		No. of
	No.	Type	items
Nottingham (2-part) Health Profile (NHP)	6	Physical mobility, pain, sleep, energy, social isolation, emotional reactions	38
	7	Employment, social life, household work, home life, sex life, interests and hobbies, holidays	7
			$\overline{45}$
Rand Health Insurance Study (HIS)	4	Physical and role functioning, mental health, social contacts, general health perceptions	25 38 3 22 $\overline{88}$
Sickness Impact Profile (SIP)	12	Sleep and rest, work, eating, mobility, home management, recreation and pastimes, body care and movement, ambulation, social interaction, alertness behaviour, emotional behaviour, communication	136
McMaster Health Index Questionnaire (MHIQ)	3	Physical function, emotional function, social function	59
Duke-UNC Profile	4	Symptom status, physical function, emotional function, social function	9 23 26 5 $\overline{63}$

quate for decision-makers who are interested in evaluating the relative impact of a treatment or intervention on different components of health and quality of life. Using a battery or profile of different health measures or a profile of interrelated components is an important alternative strategy to the aggregated index that must be considered.

Selection of Instrument

Jette [5] assessed ten indices and suggested that three major criteria need to be considered:

1. Conceptual focus, which concerns the combination of one or more of
   a) signs, i.e. directly observable events;
   b) symptoms, i.e. phenomena experienced by the individual; and
   c) performance – the capacity or actual level of function of the individual
2. Discriminatory ability, which assesses the sensitivity and specificity of items included in the indices
3. Coverage of reliability and concurrent and construct validity

Jette's table (Table 16.5) is reproduced here and it is important to note the lack of discriminatory ability of most of the measures reviewed. It is on the basis of these criteria that the Nottingham Health Profile and the Sickness Impact Profile are further described and their application in assessing quality of life in patients with angina discussed.

Nottingham Health Profile (NHP)

To obtain an individual's perception of his or her health-related problems, to elimi-nate observer bias and to prevent undue cost due to the use of physicians' time during clinical evaluation, it is important for any quality of life instrument to be self-administered. In this respect it has been claimed that the NHP is one of the most suitable instruments to assess outcome from the treatment of disease. The NHP comprises two parts: Part 1 consists of 38 questions describing health problems in terms of energy, sleep, pain, physical mobility, social isolation and emotions. Within each section of Part 1 the questions have been weighted according to perceived severity and can be combined to provide a profile of six scores to represent quality of life. Part 2 lists a number of areas which could be affected by health problems, i.e. job, household management, family life, sex life, social life, holidays and hobbies. Prelimi-nary studies have been carried out in groups of elderly patients with various clinical conditions, such as osteoarthritis, chronic bronchitis or circulatory disorders. It was also successfully tested in patients consulting their general practitioners – firemen, mine rescue workers, pregnant women, patients undergoing minor surgery and fracture victims. These studies have established the value of the NHP as a quality of life measure, independent of disease or population, and have indicated that it may have a role in the evaluation of therapeutic intervention (For key references see [4]).

Sickness Impact Profile

The Sickness Impact Profile was constructed to determine how much social factors were affected by disease. As behaviour was regarded as the best indicator of quality of life, this instrument was developed to be independent of diagnosis. It was intended that the instrument should be administered by trained interviewers but it can also successfully be self-administered. It consists of 12 categories or groups of statements on sleep and rest, emotional behaviour, body care and movement, household man-agement, mobility, social interaction, ambulation, alertness behaviour, communica-tion, work, recreation and pastimes, and eating. The categories of body care and

Table 16.5. Summary of the conceptual focus, coverage of reliability and validity, and discriminatory ability of ten health evaluation indices (Jette [5])

Health evaluation indicators	Conceptual focus			Discriminatory ability		Coverage of		
	Signs	Symptoms	Performance	Measurement level precision	Specificity of activities	Reliability	Concurrent validity	Construct validity
SIP: Bergner et al. (1)	No	Partial	Yes	Low	Low	+	+	±
IWB: Patrick et al. (2)	Yes	Yes	Yes	Low-medium	Low	+	–	±
HIS: Stewart et al. (3)	No	No	Yes	Low	Low	+	+	–
McMaster University: Sackett et al. (4)	No	Partial	Yes	Low	Low	+	+	–
FLS: Berdit and Williamson (5)	No	No	Yes	Low	Low	–	+	–
Pulses: Moskowitz and McCann (6)	No	Partial	Yes	Medium	Low	–	+	–
Barthel Index:								
Mahoney and Barthel (7)	No	No	Yes	Medium	Low	–	+	–
Convery et al. (8)	No	No	Yes	Low	Medium	+	+	–
Katz et al. (9)	No	No	Yes	Medium-high	Low	+	±	–
PGAP: Jette and Deniston (10)	No	Partial	Yes	Medium-high	High	+	±	–

+ Adequate coverage reported; – Inadequate or no coverage reported; ± Partial coverage reported

References: (1) Bergner M, Bobbitt RA, Dressel S et al (1976) The Sickness Impact Profile: conceptual formulation and methodology for the development of a health status measure. Int J Hlth Serv 6:393–415. (2) Patrick DL, Bush JW, Chen M (1973) Toward an operational definition of health. J Hlth Soc Beh 14:7. (3) Stewart AL, Ware JE, Brook RH et al (1978) Conceptualization and Measurement of Health for Adults in the Health Insurance Study. R-1987-HEW. Vols 1–8 Santa Monica: Rand Corporation. (4) Sackett DL, Chambers LW, MacPherson AS et al (1977) The development and application of indices of health: general methods and a summary of results. Am J Pub Hlth 67:423–428. (5) Berdit M, Williamson W (1973) Function limitation scale for measuring health outcomes. In: Health Status Indexes. Berg RL (Ed). Chicago: Hospital Research and Educational Trust. (6) Moskowitz E, McCann CB (1957) Classification of disability in the chronically ill and aging. J Chron Dis 5:342–346. (7) Mahoney FI, Barthel DW (1965) Functional evaluation: the Barthel Index. Md St Med J 14:61–70. (8) Convery FR, Minetter MA, Armiel D et al (1977) Polyarticular disability: a functional assessment. Arch Phys Med Rehabil 58:494–499. (9) Katz S, Downs TD, Cash HR (1970) Progress in development of an index of an ADL. Gerontologist 10:20–30. (10) Jette AM, Deniston OL (1978) Inter-observer reliability of a functional status assessment instrument. J Chron Dis 31:573–580.

movement, mobility and ambulation can be combined into a physical dimension score while the categories emotional behaviour, social interaction, alertness behaviour and communication can be combined into a psychosocial dimension. All categories can also be combined into a single overall Sickness Impact Profile score. This instrument has been used to assess the outcome and quality of life in patients with hyperthyroidism, rheumatic arthritis, sarcoma, cardiac problems, chronic obstructive pulmonary disease, breast cancer and low back pain (For key references see [1]).

Evaluation of a Quality of Life Profile

Moriyama has identified six properties for evaluating health indices [6]:
    Regardless of how it is derived, an index of health should have certain desirable properties such as:
1. It should be meaningful and understandable;
2. It should be sensitive to variations in the phenomenon being measured;
3. The assumptions underlying the index should be theoretically justifiable and intuitively reasonable;
4. It should consist of clearly defined component parts;
5. Each component part should make an independent contribution to variations in the phenomenon being measured; and
6. The index should be derivable from data that are available or quite feasible to obtain.

    Deyo [2] suggests that the following criteria can be used to assess quality of life measures:
1. *Applicability:* The content and emphasis must be appropriate to all types of clinical evaluation and acceptable to the patients who are required to complete the questionnaire.
2. *Practicality:* The respondent and professional burden involved with collecting and processing the data must be minimal.
3. *Reliability:* There should be a high degree of stability (reproducibility) of a score between one administration of the questionnaire and another within a short space of time.
4. *Validity:* There are three categories of validity that have been described for quality of life measures, namely, content, criterion and construct:
    a) Content validation checks whether an instrument is capable of measuring quality of life in a specified population and with a specified end-point such as successful therapeutic intervention.
    b) Criterion validation determines the accuracy of measurement by comparing the quality of life measure with another of proven validity.
    c) Construct validation measures the correlation of the quality of life measure with a gold standard, i.e. the best available measure of quality of life such as a clinical measure of mobility. Perfect correlation is not possible but correlation of magnitude and direction is suggested to be a useful assessment of construct validity.

5. *Specificity:* This is related to content and construct validity and assesses the ability of the quality of life measure to identify populations correctly, i. e. a specific range of scores would be expected for fit individuals which would be different to the scores for an unhealthy population.

6. *Sensitivity:* This describes the accuracy of a measurement of expected differences and the changes in quality of life as they occur in an individual. These will be due either to disease type, severity and changes in severity as a result of treatment or to progression of the disease. It is very difficult to make absolute judgements about sensitivity because, although quality of life is related to clinical aspects, it varies due to many non-medical factors. However, a quality of life measure must be capable of detecting these changes if it is to yield information useful for clinical decision-making.

Quality of Life in Angina of Assessment

Angina pectoris describes a classical syndrome characterised by chest pain and discomfort due to transient reversible myocardial ischaemia. Two types of angina are recognised; the first due to a narrowing of the major coronary arteries by atheroma; and the second in patients who have apparently normal arteries, at least as demonstrated by coronary arteriography, but exhibit the symptoms of angina. Both types lead to an imbalance between myocardial oxygen demand and supply with resulting cardiac pain. The aims of treatment include pain relief during attacks and the prevention of further cardiac complications such as myocardial infarction.

Diagnosis of angina can usually be made from a patient's description of the symptoms but sometimes features are atypical or the individual's power of expression inadequate. The patient's own report is nevertheless regarded as useful to corroborate an electrocardiograph in establishing a diagnosis. Physical examination by the doctor, on the other hand, frequently yields negative results and a careful history should be taken from the patient to establish the site of pain as well as its duration and relationship to exercise and daily activities. There are, however, many limitations to the use of physician assessments and ECG stress tests in judgements about disease severity and the success of treatment. For example, there is often a lack of correlation between the ECG exercise stress tests and the severity of angina. Also, physical and psychological factors make an important contribution to the overall severity of the angina so that a reduction in pain could be secondary to a temporary reduction in the level of daily activities or psychological factors rather than due to a direct effect on the myocardial ischaemia.

In an attempt to reproduce daily exertional capacity in the laboratory, a treadmill or bicycle ergometer is often combined with physical examination. However, this does not necessarily reflect the situation in a patient's daily life. A suitable appraisable of the patient's normal exertional capacity and pain severity by means of diary cards and a questionnaire could provide information important to the treatment and overall care of the patient. This type of evaluation is generally omitted during routine therapy and, apart from symptom diary cards, no such angina-specific questionnaires exist.

Undoubtedly, there are a number of factors which affect a patient's angina (Fig. 16.5) such as worry about his coronary artery disease, anxiety about the future, and

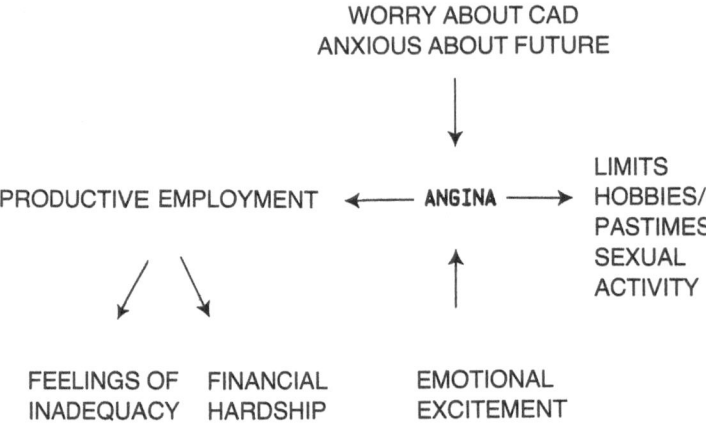

Fig. 16.5. Factors affecting angina and the effects of angina on a patient's life

emotional excitement. At the same time there is the effect that the angina will have on a patient's hobbies and pastimes, sexual activity, productive employment etc. When a patient's employment is affected, this leads to feelings of inadequacy and maybe financial hardship which, in turn, will exacerbate the angina. In the light of the problems associated with the clinical assessment of angina it is suggested that a measure of the degree of disability and psychological characteristics of the patient could be a useful means of evaluating the effects of angina and its treatment.

VandenBurg [11] has pointed out that measurement of the efficacy of anginal medication is often restricted to simple clinical measures such as the number of attacks of angina and glyceryl trinitrate (GTN) consumption or more complex laboratory measures such as exercise testing. However, the simple clinical measures are totally inadequate as a true measure of disease because patients alter their life-style, either consciously or subconsciously, to prevent the occurrence of pain. They may experience the same number of attacks of pain but may be doing more exercise or less exercise. Also, patients relate the intensity of a single attack of pain as the most important parameter of their angina and this is often not recognised. Self-assessment by patients is able to document the effects of treatment and disease on physical, social and psychosocial parameters and is a useful way to measure total health. However, there is no global measure of dysfunction to evaluate the patient with angina. Vanden-Burg [11] has assessed several visual analogue scales but Fig. 16.6 shows the problems in obtaining the answer to a simple question such as "How bad has your heart pain been?" – with 53% of patients stating that they use the severity of an individual attack to determine their response, 26% the number of attacks, 9% the duration of an attack and 6% a combination of these factors. In contrast, in answer to the question "How much trouble to you has your heart disease been?" – 39% chose frequency, 28% severity, 17% duration and 13% a combination of these factors.

In determining whether a specific instrument will be useful in evaluating health status in angina patients, it is important to ask a number of key questions such as:

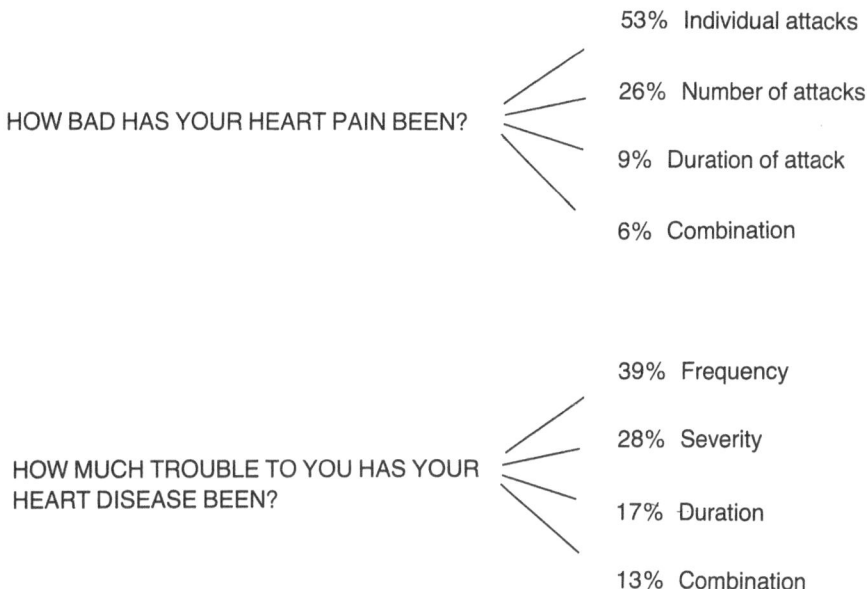

Fig. 16.6. Self-administered questionnaires in angina (VandenBurg [11])

1. Is the health measure relevant to the study question?
2. Is the health attribute capable of operational definition?
3. Is the health measure credible in terms of comprehensibility and acceptance to the patient?
4. What is the accuracy, reliability and responsiveness of the health measure to change?
5. What is the availability and cost of the health measure data source?

   In the absence of a widely accepted instrument for the measurement of quality of life, Salek et al. [8] selected and evaluated two general health questionnaires, namely, the NHP and the SIP, for their usefulness in assessing the impact that myocardial ischaemia has upon the patient with angina compared with a self-assessed global rating of the patient's health status. A total of 50 consecutive patients with an out-patient clinic appointment who presented with angina pectoris as the chief complaint were recruited for the study. Included were patients suffering from angina who had a history of angina pain with different disease severity and who had received various treatment regimens. All patients had undergone a stress test during the 6 months immediately prior to the study. All patients completed the NHP and the SIP, the order of completion (20–25 min) being randomised. In addition, the overall total well-being of the patient was self-assessed using a global rating scale (very good, good, fair, poor or very poor). The patients had their health rated on a similar scale by the physician, who also made an assessment of the severity of the angina (mild, moderate or severe).

Table 16.6. Correlations between patient-rated overall health, doctor-rated overall health, disease severity, ECG stress tests and the NHP and SIP categories

| Profiles, categories | One-way analysis of variance | | | |
	Patient-rated overall health (P value)	Doctor-rated overall health (P value)	Disease severity (P value)	ECG stress test (P value)
NHP Part 1				
Energy	NS	NS	< 0.05	NS
Pain	NS	NS	NS	NS
Emotional reaction	0.007	NS	NS	NS
Social isolation	NS	NS	NS	NS
Sleep	0.008	NS	NS	NS
Physical mobility	0.007	NS	NS	NS
NHP Part II	0.003	NS	NS	NS
Overall SIP	0.001	NS	NS	NS
Physical dimension	0.001	NS	NS	0.001
Psychosocial dimension	0.02	NS	NS	NS
Sleep and rest	0.001	NS	NS	< 0.05
Emotional behaviour	0.09	NS	NS	NS
Social interaction	0.01	NS	NS	NS

Nottingham Health Profile

Patients' assessment of their own health was consistently and significantly worse than doctor-rated patients' health. There was no correlation between doctor-rated health and the NHP scores, disease severity and the NHP scores, or ECG stress tests and the NHP scores. However, there were statistically significant correlations between patient-rated overall health and the NHP scores for emotional reaction, sleep and physical mobility and the NHP Part 2 scores (Table 16.6).

Sickness Impact Profile

Again, there was no statistically significant correlation between disease severity and SIP scores, doctor-rated health and SIP scores or overall SIP scores and ECG stress test results. However, all categories, dimensions and overall SIP scores correlated significantly with patient-rated overall health (Table 16.6).

The results from these studies demonstrate that both the SIP and the NHP can be used successfully and provide relevant information in assessing a patient population with angina. These data indicate that substantial functional impairment, psychosocial as well as physical, is associated with angina pectoris patients and both the SIP and NHP are capable of identifying these effects. Patients' overall health significantly correlated to all the SIP and NHP categories whereas doctor-rated health and disease severity failed to show such a relationship.

Conclusions

1. In angina it is important to assess the effect that disease and its subsequent treatment may have on patients' life-styles as there is substantial physical, social and psychosocial impairment.
2. Simple clinical measurements such as number of attacks, GTN consumption and exercise stress tests have severe limitations in determining the impact that the disease has on a patient's life-style.
3. Both the SIP and the NHP results correlate well with patients' global assessment of health.
4. Physicians' assessment of disease severity, their global assessment of patients' health and the results of stress tests bore little relationship to patients' assessment of their overall health.

Acknowledgments. I would like to record my appreciation to Dr. Malcolm VandenBurg of Romford Cardiovascular Research for the opportunity to collaborate in studying quality of life in patients with angina, and to Mr. Sam Salek for allowing me to illustrate some points in the review by utilising data that he obtained in a comparative study of the Sickness Impact Profile and the Nottingham Health Profile.

References

1. Bergner M (1987) Sickness impact profile. In: Walker SR, Rosser RM (eds) Quality of life – assessment and application. MTP, Lancaster
2. Deyo RA (1984) Measuring functional outcomes and therapeutic trends for chronic disease. Controlled Clin Trials 5:223–240
3. Fries JF (1983) The assessment of disability. from first to future principles. Br J Rheumatol 22 [3:Suppl]:48–58
4. Hunt SM, McEwen J, McKenna SP (1986) Measuring health status. Groom Helm, London
5. Jette AM (1980) Health status indicators: their utility in chronic disease evaluated research. J Chron Dis 37:567–579
6. Moriyama IW (1968) Problems in the measurement of health status. In: Sheldon EB, Moore WE (eds) Indicators of social change. New York
7. Patrick DL, Erickson P (1987) Assessing health-related quality of life for clinical decision making. In: Walker SR, Rosser RM (eds) Quality of life – assessment and application. MTP, Lancaster
8. Salek MS, Luscombe DK, Walker SR, Wiseman WT, VandenBurg MJ (1987) A comparison of two self-administered health profiles in angina. Lancet (submitted)
9. Stevens JC, Poston JW, Walker SR (1986) Measuring the quality of life. Proc Guild Hosp Pharmacists 22:39–49
10. Teeling Smith G (1985) Measurement of health. Office of Health Economics, London
11. VandenBurg MJ (1987) Measuring quality of life in angina. In: Walker SR, Rosser RM (eds) Quality of life – assessment and application. MTP, Lancaster
12. Ware JE, Brook RH, Davies-Avery A, Lohv KN (1981) Choosing measures of health status for individuals in general populations. Am Public Health 71:620–625

17. Quality of Life in Angina Pectoris: A Swedish Randomized Crossover Comparison Between Transiderm Nitro and Long-Acting Oral Nitrates

B. Jönsson, S. Björk, S. Hofvendahl, and J.-E. Levin

Introduction

New medical technology aims at improving quantity and quality of life. With the growing importance of chronic disease the need to develop a quantifiable definition of health-related quality of life, or health status, has become evident. Traditional biomedical parameters are often insufficient or irrelevant for assessing improvement in function.

One important application of health status measurement is in the economic appraisal of medical technology for resource allocation decisions. We are increasingly aware that it is not possible to allocate resources to all potentially beneficial medical technologies. We have to choose those which give most health for the money. Economic evaluations which take into account both costs and health benefits aim at improving decision-making in this difficult field.

Health-related quality of life can be defined as a person's subjective perception of his or her physical, psychological, and social health. This focus on subjective values is very much in line with the general development in evaluation of medical interventions that the patient's preferences should be given more weight. In economic evaluation of health care programmes this is seen as a shift from traditional cost-effectiveness to cost-utility analysis [1]. Utility refers to the value or worth of a specific level of health status and can be measured by the preferences of individuals of society for any particular set of health outcomes.

Aim of the Study

The aim of the study was to compare the patient's quality of life on a transdermal therapy 5–10 mg/24 h applied once daily with that of oral slow-release nitrates in stable angina pectoris. Secondary aims were to study "willingness to pay" (WTP) and preference of treatment in relation to the patient's quality of life.

The patch used in the trial was Transiderm-Nitro, a therapeutic system delivering nitroglycerin through the skin. The system consists of a self-adhesive nitroglycerin depot that is changed once daily. In pharmacokinetic studies it has been shown to deliver nitroglycerin at a constant rate over 24 h [2, 3].

The patch has been shown to be effective in angina pectoris in several well-controlled studies [4–11], as well as in large field studies [12, 13]. However, some studies have raised doubts about the long-term efficacy of the patch [14, 15]. The effect assessed by exercise testing seems in some patients to wear off after approximately 24 h. The contradiction between the results from the large field studies showing a high proportion of satisfied patients and some of the smaller exercise test studies can hardly be explained by the methodology alone. This provides the rationale for further studies of the effectiveness of treatment. Our hypothesis is that transdermal therapy produces a different outcome in terms of quality of life.

Measurement of Quality of Life in Clinical Trials

A number of general quality of life instruments have been developed and tested [16–19]. Both the Sickness Impact Profile (SIP) and the Nottingham Health Profile (NHP) have been used in studies of angina patients; however, at the start of the study we had no evidence whether these instruments could detect small but clinically important changes. Therefore, we looked for a disease-specific instrument as a complement to the general instrument. A specific instrument for assessing quality of life in patients with angina pectoris has been developed and tested by Wiklund and Lindvall in Gothenburg [20]. Through their cooperation we were able to use this instrument in the study. This instrument has 27 items and uses a combination of visual analog scales and Likert scales with six alternatives. The questions asked relate to the status of the person during the last week.

Even though the Wiklund–Lindvall instrument was specifically developed for angina patients, there was still uncertainty about the responsiveness of the instrument to differences in the outcome of the two treatments under investigation. We therefore decided to construct a new instrument specifically designed for the trial. Methodological standards for health status measurement have been developed [21]. When constructing and validating a quality of life instrument it is important to first specify the purpose of the instrument. Since our instrument was to be used for evaluation and not discrimination or prediction, the main issue was responsiveness. We selected 17 items which were assumed to be related to change in health status and responsive to clinically significant change. However, knowing the methodological recommendations, we had to start the trial without establishing the reproducibility, responsiveness, or validity of our new instrument. For evidence that this is not uncommon as well as a justification for the approach, see Guyatt et al. [22, p. 317].

No direct utility measurement was performed in the study. Instead the patient's preferences and WTP for treatment was investigated. Hypothetical questions about WTP can be assumed to have low validity. However, one way to test this is to investigate how health status and WTP are interrelated.

The number of anginal attacks and the consumption of sublinguals were recorded in the study, providing a further opportunity to test the validity of quality of life and WTP measures.

Method and Patients

The design of this multicenter, open, randomized, crossover study is shown in Fig. 17.1. After a 1-week run-in period, the patients were randomly divided into two groups. One group received the transdermal therapy while the other group received oral nitrates. The patients started at the lowest possible dose or the dose the patient had before randomization. All other angina pectoris therapy was kept constant during the study. After 4 weeks the treatments were switched.

One hundred and seventy-eight patients (135 men and 43 women), mean age 63 years, with stable angina pectoris were included in the study. The mean duration of angina pectoris was 6.5 years. Sixteen centers participated in the study. The number of patients per center ranged from 6 to 17.

The criteria for inclusion were definite angina pectoris according to the WHO classification, and stability for at least 3 months. The patients also had to be able to fill in the quality of life questionnaires.

Patients with recent myocardial infarction (within the last 3 months) or with hypotension and patients unable to cooperate were excluded. Also excluded were patients who had been on both oral and transdermal nitrate therapy before the study.

One hundred and sixty-six patients (93%) had been treated with long-acting nitrates before the study, the average dose of oral nitroglycerin being 9 mg/day and that of isosorbide nitrate 26 mg/day. Ninety-seven of the patients (54%) were on beta-receptor blocking agents and 65 patients (37%) were on calcium antagonists. Seventy-four (42%) of the patients in the study had suffered from at least one myocardial infarction and 36 (20%) had a diagnosis of hypertension.

The sequence of control visits is illustrated in Fig. 17.1. At each of these visits the patients were asked about the quality of life according to three different self-adminis-

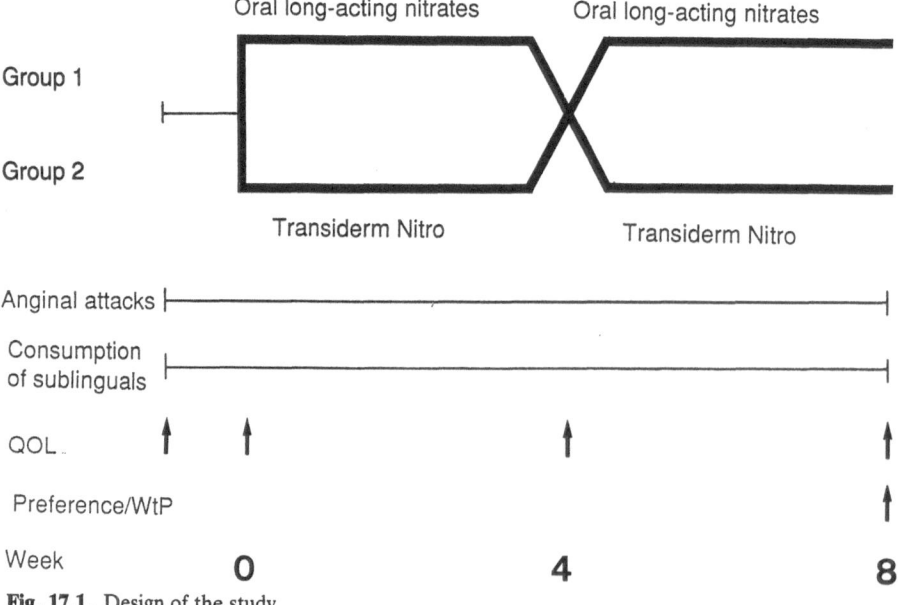

Fig. 17.1. Design of the study

tered questionnaires: SIP (Sickness Impact Profile), the Wiklund and Lindvall Angina Pectoris Quality of Life questionnaire and the CMT (Center for Medical Technology Assessment) questionnaire.

The patients were also asked to keep a diary noting their anginal attacks and their consumption of sublingual nitrate during the entire study. At the end of the study the patients were asked which nitrate treatment they preferred. They were also asked in a self-administered questionnaire about their willingness to pay for the treatment.

Statistics

The statistical analysis of data in this study has mostly been performed with non-parametric tests. The decision to restrict the use of parametric tests is based on two factors. First, most of the dependent variables are to be considered as ordinal-level data. Second, the distributions are generally nonnormal. The consequence of using nonparametric methods is a slight loss of power in detecting significant results.

The Sickness Impact Profile (SIP) is generally considered to be an interval scale questionnaire [23, 24]. Since the translation used in this study has not been scaled, we have used the weights from the English version. This must be taken into account when choosing statistical tests. We have accepted the weights as a fair estimation in the Swedish version, but we do not think this means we have an interval scale. We assume ordinal scale for the SIP.

The angina pectoris questionnaire of Wiklund and Lindvall has been thoroughly developed and tested. Several tests of reliability and homogeneity have been performed on angina patients. However, the constructors themselves recommend the use of nonparametric statistics in this case too. The CMT questionnaire is a "trial designed" questionnaire and we have decided to use nonparametric statistics in this case as well.

Some of the variables are considered to be on interval level. These variables are age, time with angina, attack rate, and consumption of sublinguals. They have been analyzed with the parametric t-test, which is appropriate in this case.

When testing dependent samples (repeated measures in one group) we have used Wilcoxon's sign-ranks test. The independent (between groups) tests have been performed with the Mann-Whitney U-test. We have chosen these because they are considered to be the most powerful nonparametric test of hypotheses of mean differences. There is no great loss in power compared with parametric mean tests.

Both methods test the difference between distribution of ranks. In the dependent case, the difference between measurements is tested, and in the independent case, the difference between groups.

Levels of significance were 5%, 1%, and 0.1%. The interpretation of differences in nonparametric tests was based on the mean rank values for measurements and groups respectively.

Due to the large amount of variables and the crossover design, many tests have been performed. This raises the problem of mass significance. This is, to some extent, avoided by using two-tailed P values, even when one-tailed hypotheses are tested. Furthermore, our conclusions were based on results where the outcome of several measures were pointing in the same direction. That is, if only single parameters showed significant results in an analysis, no specific conclusions were drawn. The amount of missing data was very small.

Computer registration of the questionnaires was performed on a personal computer using a data registration programme tailored for the study. Data were then transferred to a DEC-20 computer and the statistics were computed using a standard statistical package, SPSS[x].

Results

The two randomization groups (91 and 87 patients, respectively) were comparable with regard to age, duration of angina pectoris, treatment of angina pectoris, consumption of sublingual nitroglycerin, and frequency of anginal attacks. They did not differ in the prevalence of other diagnoses or concomitant medication (Table 17.1).

The analysis of the quality of life questionnaires is divided into two parts. The first concentrates on the two randomization groups, i.e., groups 1 and 2, see Fig. 17.1. The second part concentrates on preferences, i.e., those who preferred transdermal therapy and those who preferred oral therapy. In this part the results from the WTP questionnaires are presented.

The most responsive quality of life instrument was the trial-designed, CMT questionnaire. The disease-specific questionnaire, the Wiklund and Lindvall instrument, was the least sensitive. A possible explanation for this is that the visual analog scale

Table 17.1. Comparison of randomization groups

	Group 1 Mean	Group 1 SD	Group 2 Mean	Group 2 SD
Age	63.9	±8.7	61.9	±8.8
Time with angina (years)	6.9	±6.0	6.0	±5.5

		Group 1	Group 2
Sex	Male	76	59
	Female	21	22
Sublingual nitroglycerin	Yes	91	76
	No	4	4
Beta-blockers	Yes	48	49
	No	45	27
Calcium antagonists	Yes	34	31
	No	57	44
Other relevant diagnoses	Myoc. inf.	31	34
	Hypertens.	12	15
	Myoc. + hypertension	5	4
Long-acting nitrates	Transiderm	15	11
	Nitrong	30	21
	Nitroglyn	24	22
	Sorbangil	22	22

used in the Wiklund and Lindvall instrument is less sensitive for angina pectoris patients than the Likert scale with five alternatives used in the CMT instrument. However, we cannot rule out that the choice of items, their phrasing and configurations, also influence the responsiveness of the instrument. Even if the different quality of life questionnaires vary in responsiveness, they all point in the same direction. The presentation of the results will, therefore, concentrate on the CMT questionnaire.

Comparison According to Randomization Groups

The randomization groups were compared after 4 weeks of treatment and after 8 weeks. The only significant difference ($P < 0.05$) after 4 weeks' treatment was that patients treated with oral nitrates showed better ability to make social contacts. After 8 weeks' treatment the situation changed so that the patients treated with the transdermal therapy had better ability to make social contacts, greater desire to work, and better general well-being (Table 17.2).

Table 17.2. Randomization groups after 8 weeks of treatment: changes during the last 4 weeks

Item	Net change	Group 1:	Group 2:	Net change	P value
1. Cardiac pain mitigation	+ 16.7%	+ 34.5 0 47.8 − 17.8	+ 52.1 0 32.4 − 15.5	+ 36.6%	< 0.1
2. Cardiac pain duration	+ 42%	+ 56.8 0 28.4 − 14.8	+ 67.1 0 24.3 − 8.6	+ 58.5%	< 0.1
9. Ability to make social contacts	+ 3.7%	+ 8.6 0 86.6 − 4.9	+ 22.5 0 73.2 − 4.2	+ 18.3%	< 0.05
10. Social life in general	+ 3.6%	+ 9.6 0 84.3 − 6.0	+ 19.7 0 76.1 − 4.2	+ 15.5%	< 0.1
12. Desire to work	+ 11.8%	+ 23.6 0 64.6 − 11.8	+ 38.6 0 52.8 − 8.6	+ 30%	< 0.05
15. Ability to lead an active life	+ 7.2%	+ 19.1 0 69.0 − 11.9	+ 36.2 0 47.9 − 15.9	+ 20.3%	< 0.1
16. Possibilities to live a meaningful life	+ 7.1%	+ 16.7 0 73.8 − 9.6	+ 27.5 0 63.8 − 8.7	+ 18.8%	< 0.1
17. General well-being	+ 18.6%	+ 33.7 0 51.2 − 15.1	+ 50.7 0 33.8 − 15.5	+ 35.2%	< 0.05

+ better
0 no change
− worse

Table 17.3. Anginal attacks and consumption of sublingual nitroglycerin; comparison between the transdermal therapy and the oral therapy

	Baseline	Transiderm Nitro	Oral nitrates
Anginal attacks	0.94 ± 1.14	0.68 ± 1.21	0.68 ± 1.13
	($n = 171$)	($n = 169$)	($n = 170$)
Consumption of sublingual nitrates	0.78 ± 1.06	0.55 ± 0.88	0.65 ± 1.11
	($n = 171$)	($n = 168$)	($n = 170$)

The tables presenting the results from the quality of life instrument only report items that have a P value under 0.1. The change is presented as the positive or negative net change as well as a reduced Likert scale (from five alternatives to three) for each group. The P value shows the degree of significance between the net change in the two groups.

The results indicate that patients who started on oral long-acting nitrates and were switched to transdermal therapy had a higher quality of life than the group that had the treatments in reverse order. However, the tendency is weak and the main conclusion must be that we found no significant difference in quality of life between the two randomization groups. We tested whether the order in which the patients received the two treatments affected the outcome, but did not find this to be the case.

The number of anginal attacks and the sublingual nitroglycerin consumption during the study are presented in Table 17.2. No significant differences were found between oral nitrates and patch treatment.

Comparison According to Preferences

When asked about preference, 164 patients (92%) had a preference. Eighty of these patients (49%) preferred transdermal therapy and 84 (51%) preferred oral therapy.

Table 17.4. Anginal attacks and consumption of sublingual nitroglycerin during the last 4 weeks; comparison between the transdermal therapy and the oral therapy among patients preferring the transdermal therapy

Week	Oral long-acting nitrates		Transiderm-Nitro		
	Mean	SD	Mean	SD	P value
Anginal attacks/day					
1	1.0	1.4	0.4	0.6	≤ 0.05
2	1.0	1.4	0.5	0.7	NS
3	1.0	1.5	0.4	0.6	≤ 0.05
4	0.9	1.4	0.5	0.7	NS
Consumption/day of sublingual nitrates					
1	1.0	1.7	0.3	0.5	≤ 0.05
2	1.0	1.8	0.4	0.6	≤ 0.05
3	1.1	1.9	0.4	0.6	≤ 0.05
4	1.1	1.8	0.4	0.7	≤ 0.05

Table 17.4 shows the number of anginal attacks and the sublingual nitroglycerin consumption during the last 4 weeks of treatment for those who preferred the transdermal therapy and were treated with this therapy compared with those who were treated with oral therapy. The results show a significantly lower sublingual nitroglycerin consumption and also a reduced attack rate during weeks 1 and 3. No significant differences were found for those patients who preferred oral nitrates.

The results of the quality of life assessments according to preference are shown in Tables 17.5–17.9. In Tables 17.5 and 17.6 the preference groups are compared during the last 4 weeks of treatment. Those patients who preferred the transdermal therapy and were treated with this therapy are compared with those who preferred transder-

Table 17.5. Findings in patients preferring transdermal therapy during the last 4 weeks of treatment

Item	Net change	Oral treatment	Transdermal treatment	Net change	P value
1. Cardiac pain mitigation	− 10.9%	+ 21.6 0 45.9 − 32.5	+ 65.9 0 24.3 − 9.8	+ 56.1%	< 0.0001
2. Cardiac pain duration	+ 11.2%	+ 38.9 0 33.4 − 27.7	+ 80.5 0 17.1 − 2.4	+ 78.1%	< 0.0001
3. Cardiac pain frequency	+ 8.4%	+ 38.9 0 30.6 − 30.5	+ 73.2 0 19.5 − 7.3	+ 65.9%	< 0.0005
4. Headache	− 8.3%	+ 16.7 0 58.3 − 25.0	+ 64.7 0 23.5 − 11.8	+ 52.9%	< 0.05
6. Acting mobility	− 2.9%	+ 11.8 0 73.5 − 14.7	+ 43.2 0 40.6 − 16.2	+ 27%	< 0.05
7. Physical health	− 11.7%	+ 17.7 0 52.9 − 29.4	+ 45.9 0 48.7 − 5.4	+ 40.5%	< 0.005
9. Ability to make social contacts	− 3%	+ 6.1 0 84.8 − 9.1	+ 26.8 0 70.8 − 2.4	+ 24.4%	< 0.01
10. Social life in general	± 0%	+ 8.8 0 82.4 − 8.8	+ 24.4 0 73.2 − 2.4	+ 22%	< 0.05
11. Ability to work	− 2.9%	+ 11.4 0 74.3 − 14.3	+ 41.4 0 41.6 − 17.0	+ 24.4%	< 0.1
12. Desire to work	± 0%	+ 14.3 0 71.4 − 14.3	+ 47.5 0 47.5 − 5.0	+ 42.5%	< 0.005
17. General well-being	− 8.5%	+ 17.2 0 57.1 − 25.7	+ 61.0 0 29.3 − 9.7	+ 51.3%	< 0.0005

Table 17.6. Findings in patients preferring oral nitrates during the last 4 weeks

Item	net change	Oral treatment	Transdermal treatment	Net change	P value
1. Cardiac pain mitigation	+ 35.9%	+ 43.4 0 49.1 - 7.5	+ 33.3 0 43.4 - 23.3	+ 10%	< 0.05
2. Cardiac pain duration	+ 63.4%	+ 69.2 0 25.0 - 5.8	+ 48.3 0 34.5 - 17.2	+ 31.1%	< 0.05
3. Cardiac pain frequency	+ 52.9%	+ 60.4 0 32.1 - 7.5	+ 48.3 0 31.0 - 20.7	+ 27.6%	< 0.05
4. Headache	+ 31.3%	+ 37.6 0 56.3 - 6.3	+ 21.4 0 14.3 - 64.3	- 42.9%	< 0.05
7. Physical health	+ 33.3%	+ 37.5 0 58.3 - 4.2	+ 20.6 0 65.5 - 13.8	+ 6.8%	< 0.1

mal therapy and were treated with oral therapy (Table 17.5). A similar comparison was made for those who preferred oral nitrates (Table 17.6).

Table 17.5 shows that the patients who preferred the transdermal therapy significantly improved on transdermal therapy as regards headache, acting mobility, physical health, ability to make social contacts, social life in general, and desire to work. The significance is particularly high ($P < 0.001$) for mitigation, duration, and frequency of cardiac pain and general well-being.

Table 17.6 shows that the patients who preferred oral nitrates significantly improved on oral treatment as regards mitigation, duration, and frequency of cardiac pain as well as headache.

In Tables 17.7 and 17.8 transdermal therapy is compared with oral therapy for patients who preferred transdermal therapy (Table 17.7) and for patients who preferred oral therapy (Table 17.8).

Table 17.7 shows that transdermal therapy is favored on most of the items. The significance is also strong. Mitigation, duration, and frequency of cardiac pain, acting mobility, and general well-being show the highest statistical significance ($P < 0.005$).

When Tables 17.7 and 17.8 are compared, there is a striking difference. Only four items in Table 17.8 have a significant difference, compared with eight in Table 17.7. The degree of significance is also higher in Table 17.7 than in Table 17.8. However, when findings in transdermal-preferring patients on transdermal therapy was compared to oral-preferring patients on oral therapy, the results are not as strong as for the other comparisions. This is quite natural because the two therapies are compared when they are used optimally. As shown in Table 17.9, cardiac pain mitigation, social life in general, and general well-being are favored in transdermal therapy, and the difference is statistically significant.

Table 17.7. Comparison of transdermal and oral therapy in patients preferring transdermal therapy

Item	Net change	Oral treatment	Transdermal treatment	Net change	P value
1. Cardiac pain mitigation	+ 16.3%	+ 38.8 0 38.7 − 22.5	+ 57.7 0 34.6 − 7.7	+ 50.0%	< 0.005
2. Cardiac pain duration	+ 34.1%	+ 50.6 0 32.9 − 16.5	+ 71.8 0 25.6 − 2.6	+ 69.2%	< 0.0005
3. Cardiac pain frequency	+ 20.2%	+ 46.8 0 26.6 − 26.6	+ 64.1 0 26.9 − 9.0	+ 55.1%	< 0.005
6. Acting mobility	+ 8.5%	+ 24.6 0 59.4 − 15.9	+ 41.4 0 48.6 − 10.0	+ 31.4%	< 0.005
7. Physical health	+ 11.6%	+ 31.9 0 47.8 − 20.3	+ 45.7 0 48.6 − 5.7	+ 40%	< 0.05
11. Ability to work	+ 15.4%	+ 29.5 0 56.4 − 14.1	+ 39.0 0 48.1 − 13.0	+ 26%	< 0.1
12. Desire to work	+ 18%	+ 30.8 0 56.4 − 12.8	+ 42.7 0 50.6 − 6.7	+ 36%	< 0.1
15. Ability to lead an active life	− 1.7%	+ 21.1 0 56.6 − 22.4	+ 33.8 0 52.7 − 13.5	+ 20.3%	< 0.05
16. Possibilities to live a meaningful life	+ 7.9%	+ 21.1 0 65.8 − 13.2	+ 32.9 0 56.2 − 11.0	+ 21.9%	< 0.05
17. General well-being	+ 11.6%	+ 30.8 0 50.0 − 19.2	+ 63.6 0 26.0 − 10.4	+ 53.2%	< 0.001

Willingness to Pay

After 8 weeks, the patients were asked how much they were prepared to pay for their preferred nitrate therapy. They were asked if they were prepared to pay 50, 100, 200, or 500 SEK (Swedish crowns) a month. The patients who preferred the transdermal therapy were also asked if they would be willing to pay the higher cost of this therapy out of their own pockets. The preference groups are equally distributed among those willing to pay 50 SEK and 100 SEK a month for the preferred treatment. Twice as many patients who preferred transdermal therapy to oral therapy would have paid 200 SEK (13 vs. 5), while the opposite tendency was found among those prepared to pay 500 SEK (8 vs. 10) a month for their nitrate therapy. Sixty-eight of the 80 transdermal-

Table 17.8. Comparison of transdermal and oral therapy in patients preferring oral therapy

Item	Net change	Oral treatment	Transdermal treatment	Net change	P value
2. Cardiac pain duration	+ 53%	+ 61.4 0 30.1 − 8.4	+ 51.2 0 31.7 − 17.1	+ 34.1%	< 0.05
6. Acting mobility	+ 24.7%	+ 35.1 0 54.4 − 10.4	+ 18.4 0 73.7 − 7.9	+ 10.5%	< 0.05
7. Physical health	+ 26%	+ 32.5 0 61.0 − 6.5	+ 20.0 0 65.3 − 14.7	+ 5.3%	< 0.05
11. Ability to work	+ 18.8%	+ 31.3 0 56.3 − 12.5	+ 24.1 0 57.0 − 19.0	+ 5.1%	< 0.1
17. General well-being	+ 28%	+ 37.8 0 52.4 − 9.8	+ 24.4 0 56.1 − 10.4	+ 14%	< 0.05

Table 17.9. Comparison of findings in transdermal-preferring patients on transdermal therapy and in oral-preferring patients on oral therapy

Item	Net change	Oral pref.	Transdermal pref.	Net change	P value
1. Cardiac pain mitigation	+ 26.2%	+ 36.9 0 52.4 − 10.7	+ 57.7 0 34.6 − 7.7	+ 50%	< 0.05
10. Social life in general	+ 6.3%	+ 10.1 0 86.3 − 3.8	+ 21.4 0 77.3 − 1.3	+ 20.1%	< 0.05
17. General well-being	+ 28.1%	+ 37.8 0 52.4 − 9.7	+ 63.7 0 26.0 − 10.4	+ 53.3%	< 0.005

preferring patients (85%) and 64 of the 84 oral-preferring patients (76%) answered the WTP question.

Of the patients who preferred the transdermal therapy, 46 out of 78 (59%) were willing to pay the higher cost of the transdermal therapy out of their own pockets.

The cost of the oral therapy and the cost of the transdermal therapy were compared with the WTP in the two preference groups. The WTP for the transdermal-preferring patients (Fig. 17.2) shows that between 26% and 40% of the 85% of the patients who answered the question thought the transdermal therapy was worth its price. The WTP for the oral-preferring patients (Fig. 17.3) shows that between 64% and all of the 76% who answered the question thought that the oral therapy was worth its price.

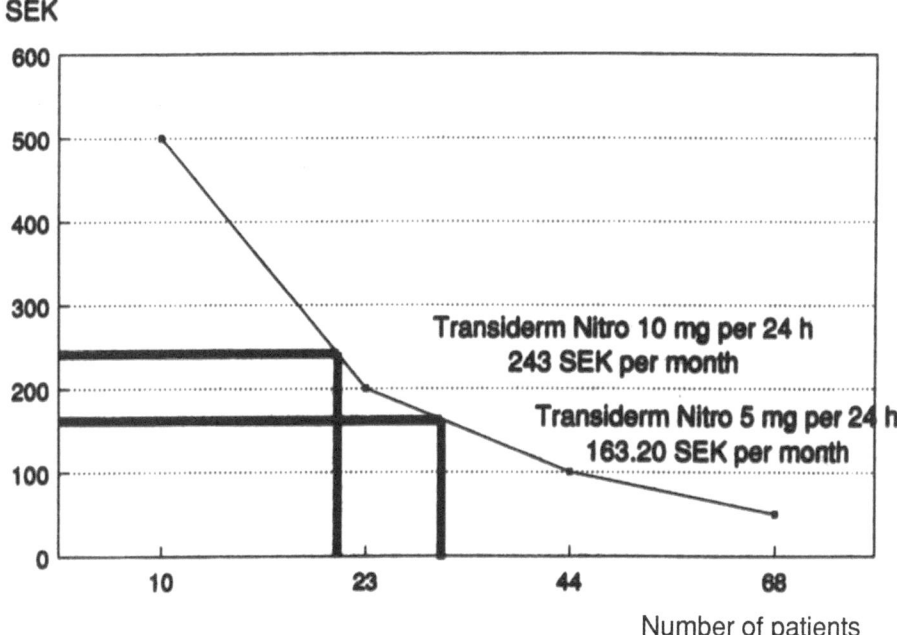

Fig. 17.2. Willingness to pay among patients preferring transdermal therapy

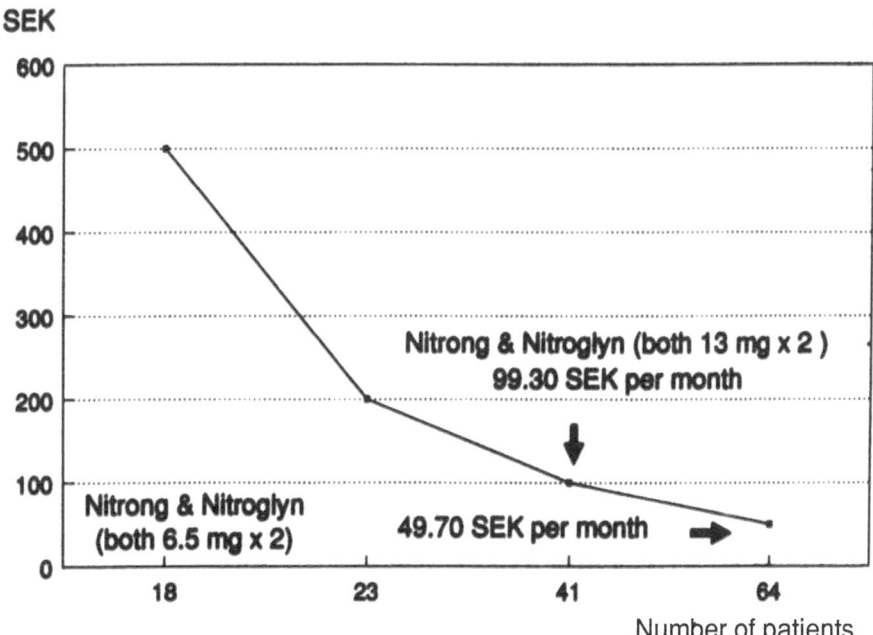

Fig. 17.3. Willingness to pay among patients preferring oral therapy

Discussion

When the results are analyzed, there appear some differences between the groups after 8 weeks treatment, i.e., when the patients had both treatments. These differences are in favor of the transdermal therapy and are significant for ability to make social contacts, desire to work, and general well-being. As there are only slight differences in the consumption of sublingual nitrates and no differences in the attack rate, the analysis does not allow any special conclusions.

However, the most interesting results are found when the preference groups are analyzed. Those who preferred the transdermal therapy had a significant increase in quality of life when receiving this therapy. These patients also displayed lower attack rates and needed less sublingual nitrates. Analogous findings (although not as strong) for oral-preferring patients were observed regarding quality of life.

The changes in attack rate and consumption of sublingual nitrates take place parallel with changes in quality of life. This supports the assumption that the quality of life instrument is valid. It also supports the assumption that the quality of life measurement presents information which the medical measurement cannot.

The transdermal-preferring group had a higher statistically significant improvement in quality of life than the oral-preferring group. However, this does not imply that a single angina pectoris patient actually improves his quality of life if he is treated with the transdermal therapy. What it does indicate is that if he belongs to the patch-preferring group his quality of life will improve on patch treatment more than if he belongs to the oral-preferring group and is on the oral treatment. The task then is to identify the two preference groups in order to decide which therapy is best suited to each patient.

According to the data in the trial, it is not possible to identify and characterize the two groups. What determines who will be a transdermal-preferring patient and who will be an oral-preferring one cannot be explained using the available data. Possibly some personality trait could help to describe the typical transdermal-preferring patient and the typical oral-preferring patient. One way to find out whether a patient is transdermal- or oral-preferring is to give him both treatments and simply ask him which therapy he prefers. However, whether this is feasible depends on financial as well as practical considerations.

Approximately 25% of the transdermal-preferring patients who answered the question (85% did) thought the higher cost (cost for oral vs. cost for patch) of the most expensive patch (243.90 SEK/month) was worth its price and 40% thought the higher cost of the cheaper patch (163.20 SEK/month) was worth its price. Sixty-four percent of the oral-preferring patients who answered the question (76% did) thought the most expensive treatment (99.30 SEK/month) was worth its price and all thought the least expensive oral therapy (49.70 SEK/month) was worth its price. These results indicate that the WTP correlates to the degree of the change in the quality of life. However, no conclusions can be verified since the question is hypothetical and there are no sanctions connected to the answers. Although it is possible to present a correlation between WTP and the degree of quality of life, the correlation cannot be taken to be more than a tendency. The number of missing cases also support this. There are two reasonable explanations for the missing cases. The first and less reasonable explanation is that some patients simply have missed the question and that this has been done unconsciously. The other is that they omitted the answer consciously, maybe because the considered the question complicated and difficult to assess in monetary terms. If the latter explanation applies and the number of missing cases in both the transdermal-preferring group and the oral-preferring group are the same, i.e., the percentages between the conscious and the unconscious are equal in the two groups, a slight tendency shows up. More transdermal-preferring than oral-preferring patients answered the WTP question (85% vs. 76%). One explanation is that the higher the degree of change in quality of life, the better the possibility of assessing what this difference means in monetary terms.

Conclusions

It was easy to motivate the physicians for the study and the compliance among the patients was good. The validity of the trial-designed quality of life instrument is demonstrated by the parallelism of attack rate and consumption of sublinguals. A tendency was also found for a correlation between quality of life and WTP.

The comparison between the randomization groups did not allow any conclusion about quality of life. An equal number of patients preferred the transdermal therapy and the oral therapy. Each preference group had an improved quality of life on its preferred therapy. When the groups were compared on their preferred therapies, differences were found. The transdermal-preferring group experienced greater improvement in quality of life than did the oral-preferring group. This supports the conclusion that transdermal therapy improves quality of life to a higher degree than does oral therapy. However, this does not indicate which therapy is the more effective for the individual angina pectoris patient if we do not know which therapy he prefers.

In this study, it was not possible to identify the criteria for the preference groups. Consequently, the patient has to try both therapies in order to decide which he prefers; the feasibility of this procedure depends on both practical and financial considerations.

The equal preferences for oral therapy and transdermal therapy support the conclusion that there is a large group of angina pectoris patients now on oral therapy who would benefit more – i.e., attain a higher quality of life – from transdermal therapy.

References

1. Torrance GW (1986) Measurement of health state utilities for economic appraisal. J Health Econ 5:1–30
2. Müller P, Imhof PR, Burkart F, Chu L-C, Gérardin A (1982) Human pharmacological studies of a new transdermal system containing nitroglycerin. Eur J Clin Pharmacol 22:473–480
3. Place VB A. The bioavailability of nitroglycerin from TTS in normal volunteers. Data in file, Ciba-Geigy
4. Terland O, Eidsaunet W (1986) A double-blind multicenter, cross-over general practice study of glyceryl trinitrate delivered by a transdermal therapeutic system in angina pectoris. Curr Ther Res 39:214–222
5. Georgopoulos AJ, Markis A, Georgiadis H (1982) Therapeutic efficacy of a new transdermal system containing nitroglycerin in patients with angina pectoris. Eur J Clin Pharmacol 22:481–485

6. Midtboe K (1985) A comparative study of Nitroderm TTS versus placebo in the treatment of stable angina pectoris. Hans Huber, Bern

7. Martines C (1984) Comparison of the prophylactic anti-anginal effect of two doses of Nitroderm TTS in out-patients with stable angina pectoris. Curr Ther Res 36:483–489

8. Dickstein K, Knutsen H (1985) A double-blind multiple cross-over trial evaluating a transdermal nitroglycerin system vs placebo. Eur Heart J 6:50–56

9. Imhof PR, Müller P, Georgopoulos AJ, Garnier B (1985) Nitroderm TTS versus oral isosorbide nitrate: a double-blind trial in patients with angina pectoris. Acta Ther 11:155–170

10. Kapoor AS, Dang NS, Reynolds RD (1985) Sustained effects of transdermal nitroglycerin in patients with angina pectoris. Clin Ther 7:674–679

11. Muiesan G, Agabiti-Rosei E, Muiesan L et al. (1986) A multicenter trial of transdermal nitroglycerin in exercise-induced angina: individual antianginal response after repeated administration. Am Heart J 112:233–235

12. Letzel H, Johnson LC (1984) Results of a multicenter field study in 37596 patients: therapy of angina pectoris with Nitroderm TTS. Med Welt 35:326–332

13. Bridgman KM, Carr M, Tattersall AB (1984) Post-marketing surveillance of the Transiderm-Nitro patch in general practice. J Int Med Res 12:40–45

14. Crean PA (1984) Failure of transdermal nitroglycerin to improve chronic stable angina: a randomised, placebo-controlled, double-blind, double cross-over trial. Am Heart J 108:1494–1500

15. Sullivan M, Savvides M, Abouantoun S, Madsen E, Froelicher V (1985) Failure of transdermal nitroglycerin to improve exercise capacity in patients with angina pectoris. J Am Coll Cardiol 5:1220–1223

16. Ware JE, Brook RH, Davies-Avery A et al. (1980) Conceptualization and measurement of health for adults in the health insurance study, vol 1: model of health and methodology. Rand Corporation, Santa Monica, Calif.

17. Kaplan RM, Bush JW, Berry CC (1976) Health status: types of validity and the index of well-being. Health Serv Res 11:478–507

18. Bergner M, Bobbit RA, Carter WB et al. (1981) The Sickness Impact Profile development and final revision of a health status measure. Med Care 19:787–805

19. Hunt SM, McKenna SP, McEwen J et al. (1980) A quantitative approach to perceived health status: a validation of study. J Epidemiol Community Health 34:281–286

20. Wiklund I (to be published) Quality of life in patients with cardiovascular diseases (in Swedish). Scand J Behav

21. Kirshner B, Guyatt G (1985) A methodological framework for assessing health indices. J Chron Dis 38:27–36

22. Guyatt G, Crowe J, McKelvie R, Runions J, Oldridge N (1986) Assessing quality of life in cardiovascular disease: a general approach and an example in patients with myocardial infarction. Quality of Life and Cardiovascular Care, Nov/Dec, pp 304–318

23. Carter WB, Bobbit RA, Bergner M (1976) Validation of an interval scaling. Health Serv Res 11:516–528

24. Gilson BS, Gilson JS, Bergner M, Bobbit RA, Kressel S, Pollard WE, Vesselago M (1975) The Sickness Impact Profile – development of an outcome measure of health care. Am J Public Health 65:1304–1310

18. Transdermal Nitrate Therapy in Coronary Heart Disease from the Patient's Point of View – A Quality of Life Study in the Federal Republic of Germany

B. GÜTHER

Introduction

Independent of the scientific discussion on so-called nitrate tolerance [2], which basicly concerns all nitrates and their various forms of administration, modern transdermal nitrate therapy has become impressively popular. The limited number of studies published in Anglo-Saxon countries [1, 4] discovered that patients suffering from coronary disease preferred transdermal nitroglycerin therapy to oral nitrate therapy. The reasons behind such a high acceptance of transdermal therapy have so far only been elucidated from certain angles. This study, conducted by Infratest Health Research in 1986, was intended to yield a more comprehensive insight into the criteria and components used for the assessment of transdermal nitrate therapy by the patients concerned. To this end, a written inquiry was carried out in the summer of 1986 with patients suffering from symptomatic heart disease and their attending physicians. The study objective was to register changes in the general condition of patients, their anginal troubles, and the psychological and social effects produced by disease and therapy after the prescription of Nitroderm TTS.

Study Design and Patient Selection

A total of 528 physicians and 1065 patients were included in the study; i. e., there were approximately two patients per doctor. The physicians included in the study were general practitioners/specialists in general medicine ($n = 385$) and internists in office practice ($n = 143$) who were selected at random from the whole Federal Republic of Germany. Qualified interviewers requested the doctors to encourage the next few patients they treated with Nitroderm TTS to participate in this study. The survey instrument used was a questionnaire which the patients were to complete and return to their physicians in a sealed enevelope, thus preventing the doctors from seeing the completed form. The physicians for their part recorded information with respect to diagnosis and drug therapy on a separate documentation sheet.

Structure of Patient Sample

Of the patients interviewed, 55% were male, with an average age of 64 years; the women participating in the study were on average aged 69. Only a small proportion of patients were still working (one-third of male and one-tenth of female respondents) (Table 18.1).

Three-quarters of patients had had more than 6 months' experience with transdermal nitrate therapy (Table 18.2).

Table 18.1. Structural data on respondents

Sex	Male	55%
	Female	45%
Age	< 60 years	27%
	60–69 years	35%
	≥ 70 years	38%

Most of the patients wore the patch 24 hours a day. Sixty-two percent of patients were prescribed transdermal nitrate therapy because other treatment had not produced satisfactory results; only a minority (16%) were prescribed the patch as primary therapy for coronary insufficiency. Almost all patients suffered not only from coronary heart disease but also from other conditions. Hypertension (54%), heart insufficiency (52%), and diabetes mellitus (25%) were the diseases cited most frequently in this context.

The data physicians made available with respect to the severity of their patients' condition showed 11% to be suffering mild angina pectoris, 59% moderate, and 29% severe.

Only a small proportion of patients had not received drug treatment for coronary insufficiency prior to the prescription of Nitroderm TTS. The nitrate patch was prescribed to only 16% of patients as primary therapy for coronary insufficiency. Approximately 70% of patients, however, had previously been treated with nitrates, beta-blockers, and/or calcium antagonists before being prescribed a nitrate patch.

Table 18.2. Average duration of medical treatment for angina pectoris and the use of Nitroderm TTS (in months)

Undergone treatment for angina pectoris for	%	User of Nitroderm TTS for	%
< 1 year	15	< 6 months	25
1–2 years	21	6–12 months	22
3–4 years	25	1–2 years	33
> 4 years	36	> 2 years	17
Don't know/no response	3	Don't know/no response	3
	100		100

Effectiveness of Transdermal Nitrate Therapy

The effectiveness of transdermal nitrate therapy and the physical condition of patients can be assessed by way of the statements made with respect to frequency, occasion, duration, and severity of anginal attacks experienced prior to and after the prescription of transdermal nitrate therapy. Additional indicators to prove the effectiveness of this therapy are the frequency with which doctors were visited and the frequency and duration of periods when patients were unable to work. By far the largest portion of patients interviewed reported that they experienced anginal troubles less frequently (Table 18.3).

Table 18.3. Frequency of anginal attacks before and since treatment with Nitroderm TTS

	Before Nitroderm TTS treatment	Since Nitroderm TTS treatment
Never	4	15
Several times a month and less frequently	27	48
At least once a week	37	26
Once a day	10	3
Several times a day	17	2

This also holds true when the severity of the disease is taken into account. Physical strain, mental strain, and cold – the three potential factors triggering anginal attacks – occurred less frequently to varying degrees (Table 18.4).

The duration of anginal attacks decreased significantly with Nitroderm TTS treatment, from a mean of 3.2 min ($n = 1020$; SD $= 1.4$) to a mean of 1.9 min ($n = 903$; SD $= 1.4$), as did the intensity or painfulness, from a rating of 3.6 ($n = 1020$; SD $= 0.9$) on a five-point scale (from hardly painful to very painful) to a rating of 2.2 ($n = 903$; SD $= 0.8$).

Table 18.4. Percentage, of patients with anginal attacks before and since treatment with Nitroderm TTS, under the influence of various factors

	Before Nitroderm TTS treatment (%)	Since Nitroderm TTS treatment (%)
Physical strain	85	78
Mental strain	69	55
Cold	26	15

A total of 81% out of 1065 patients interviewed claimed that they had experienced an alleviation of anginal symptoms since nitrate patch therapy had been initiated, whereas 17% of respondents stated that there had been no change and 1% of patients claimed that the symptoms had deteriorated. When analyzing the result according to the severity of the condition it becomes obvious that according to the overall assessment of patients, improvement was experienced by all patient groups (Table 18.5).

Table 18.5. Changes in the physical condition of patients since the commencement of Nitroderm TTS treatment

	Total	Light form of AP	Medium form of AP	Severe AP
	$n = 1060$ (%)	$n = 118$ (%)	$n = 631$ (%)	$n = 311$ (%)
Improved	81	76	81	83
Remained the same	17	20	17	16
Deteriorated	1	3	1	1

AP, angina pectoris

Nonsomatic Indicators of Effectiveness

Apart from the aspect of relief or alleviation of physical discomfort, the study aimed in particular at determining what effect the patch therapy produces with respect to quality of life, which patients can only experience subjectively. A further secondary aspect was the positive financial effect accruing from the fact that patients required medical services less frequently and that periods when they were unable to work were shorter.

Effects of Nitrate Patch Therapy with Respect to Inability to Work and Utilization of Medical Services

More than two-thirds of patients claimed that they consulted their physician less frequently for anginal troubles after the initiation of patch therapy (Table 18.6).

Of patients suffering from angina pectoris and currently working two-thirds were at least once certified by their doctor as being unable to work because of heart disease. Patients were unable to work for an average of 37 days. In severe cases patients were on average unable to work for 45 days. Of the 161 patients who reported that they had been unable to work prior to the initiation of Nitroderm TTS therapy, 88% claimed that during Nitroderm TTS therapy they had been unable to work less frequently or over shorter periods. Broken down, the figures were 75% for the mild form of angina pectoris, 92% for the moderate form, and 83% for the severe form.

Table 18.6. Visits paid to physician due to problems related to angina pectoris since the commencement of Nitroderm TTS treatment ($n = 1065$)

Higher frequency	2%
Same frequency	27%
Lower frequency	69%

Influence on Quality of Life

Since coronary heart diseases cannot be exclusively assessed in terms of clinical symptoms, the questionnaire also contained questions about the quality of life. These questions were designed to determine to what degree patients were handicapped
a) in the accomplishment of essential everyday chores, such as household chores, shopping, and professional activity, and
b) in their life-style, as manifested in social contacts within the family, with neighbors, and with friends.

On five-point scales ranging from not at all handicapped to extremely handicapped, the improvements in these two areas since Nitroderm TTS treatment were from a mean of 3.7 (SD = 1.0) to 2.1 (SD = 0.8) for point (a) and from a mean of 3.4 (SD = 1.1) to 2.1 for point (b).

The higher quality of life experienced by patients undergoing Nitroderm TTS therapy was probably due to its antianginal action providing the range of action patients needed for the performance of everyday chores and social contacts. This was true for all groups of patients analyzed by the degree of severity of their condition (Figs. 18.1, 18.2).

By way of two statement lists respondents were asked about the advantages and disadvantages perceived in connection with the nitrate patch. The disadvantages most frequently cited are presented in Table 18.7. The advantages, however, clearly outweigh the disadvantageous side-effects (Table 18.8).

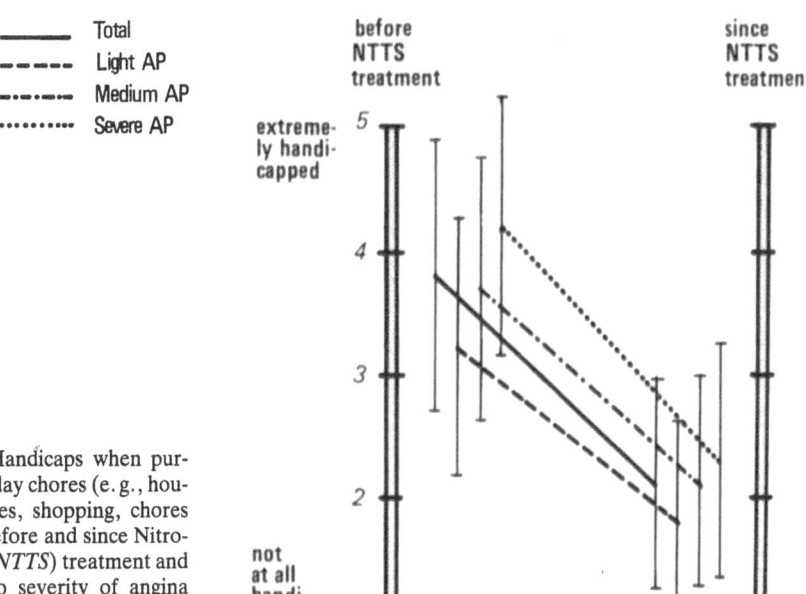

Fig. 18.1. Handicaps when pursuing everyday chores (e. g., household chores, shopping, chores at work), before and since Nitroderm TTS (*NTTS*) treatment and according to severity of angina pectoris

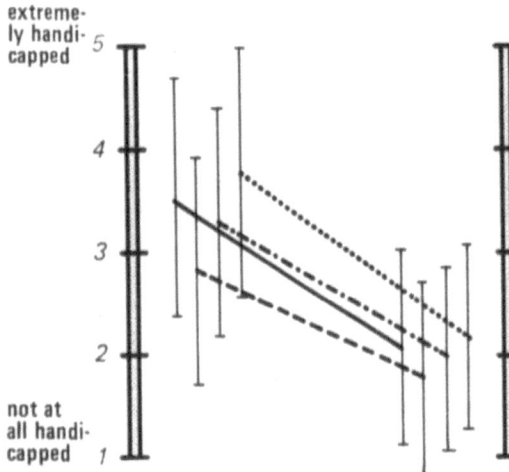

Fig. 18.2. Handicaps in connection with aspects of life-style not covered in Fig. 18.1 (e.g., family life, contacts with neighbors and friends), before and since Nitroderm TTS (*NTTS*) treatment and according to severity of angina pectoris

Table 18.7. Most frequently cited disadvantages of Nitroderm TTS treatment: 5 = applies fully; 1 = does not apply at all

Disadvantage	Average
Causes flushing	2.3
Leaves troublesome "sticky marks" on skin	2.1
Causes itching	2.0
Causes headaches	2.0

Table 18.8. Most frequently cited advantages of Nitroderm TTS treatment: 5 = applies fully; 1 = does not apply at all

Advantage	Average
Simple to use	4.7
Attacks occur less frequently	4.4
Have longer periods without troubles	4.3
Feel safer	4.3
Feel less afraid of attacks	4.2
Plaster gives me the feeling of being better looked after by doctor	4.1

Discussion

According to the assessment made by the patients interviewed, quality of life improved not only in the somatic field but also in terms of well-being and social contacts. These are probably decisive reasons for the good compliance with and acceptance of this patch. Whether the effects result from the comfortable handling, the low dosage (once a day), or psychological factors or all three cannot be finally decided from the data of this study. It may be argued that there are certain weaknesses in the study design (retrospective approach, self-selection of patients, assumed placebo effects). The central issue of the study, however, was to demonstrate that patients under Nitroderm TTS therapy show considerable improvements on several levels or dimensions of their daily life experience. Other studies with a different approach and based on different types of data show similar results [3]. So there seem to be more or less consistent effects, validated by repeated measurements, with regard to patients' somatic and general well-being.

References

1. Bridgman KM, Carr M, Tattersall AB (1984) Post-marketing surveillance of the Transiderm-Nitro patch in general practice. J Int Med Res 12:40–45
2. Cowan JC (1986) Nitrate tolerance. Int J Cardiol 12:1–19
3. Imhof PR, Vuillemin Th, Gévardin A, Racine A, Müller P, Follath F (1984) Studies of the bioavailability of nitroglycerin from a transdermal therapeutic system (Nitroderm TTS). Eur J Clin Pharmacol 27:7–12
4. Letzel H, Johnson LC (1984) Therapie der Angina pectoris mit Nitroderm TTS. Ergebnisse einer multizentrischen Untersuchung an 37596 Patienten. Med Welt 35:326–332

19. Angina Pectoris Prophylaxis and Quality of Life

H. Schneider

Introduction

Angina pectoris is a well-known symptom of coronary heart disease. Anatomically, a reduction in diameter resulting from sclerosis can be observed in one or several coronary branches. Clinically, an angina pectoris attack is characterized by acute pain not directly in the region of the heart. In some cases this pain is described as a rather weak compression or as a vague feeling of unease. Often, however, adjectives such as "suffocating," "choking," or "crushing" are used.

The constant fear of getting another attack characterizes the life of the angina pectoris sufferer. His behavior and activities are dominated by the fact that certain conditions – physical or psychological stress, low temperatures, excitement or pressure, or even just a sumptuous meal or a change in the weather – can provoke a painful attack. Therefore, he often feels obliged to rest and to adopt more careful behavior in general; he avoids any kind of physical strain when walking or working in the garden, during sports, and on vacation. This is, in turn, frequently accompanied by a decrease in self-esteem. Life seems joyless, the future looks bleak. The patient is unhappy about his life and often becomes depressed.

The primary therapeutic goal of angina pectoris treatment is a reduction in the number and intensity of angina attacks. Therefore, certain criteria are used to determine whether a given therapy is clinically effective. To determine the success of a therapy by objective parameters is a medically and scientifically accepted procedure. However, these parameters alone do not indicate to what extent a treatment relieves the patient from anxiety and allows him to perform a wide range of daily activities, thus leading to a more productive and satisfying life. Only the patient himself can assess the subjective benefits of a therapy, as opposed to its objective clinical efficacy. The final assessment of a therapy and the choice of the best specific procedure can only be made after patients personally have had the chance to evaluate that therapy and to rate its effects on well-being and on conditions of life. In other words, the patient's "quality of life" represents as important a factor in the final rating of a therapy as does its clinical effectiveness.

Drug therapies can have different effects on the quality of life of a patient, even if their clinical efficacy is comparable. Several studies document this fact impressively for therapeutic domains such as hypertension [1–6]. The aim of the trial described in this report was to examine whether different drug therapies for angina pectoris also differently affect the patient's quality of life. Two treatment modalities were chosen from the spectrum of currently available therapeutic alternatives for the prophylaxis of angina pectoris: the standard therapy with long-acting oral nitrates and the transdermal therapeutic system Nitroderm TTS, which has recently become available.

Patients

Demographic Data

One hundred and forty-seven male and female outpatients were enrolled in the trial according to the following criteria:
1. Patients with coronary heart disease (NYHA II–III) established by
   a) typical anamnesis (attacks induced only by exertion or by cold) with at least three attacks per week before preexisting treatments or
   b) an ergometric test (exercise electrocardiogram with lowering of the ST segment by at least 1 mm) or
   c) coronary angiography.
2. Patients who previously were treated with long-acting oral nitrates, alone or in combination with other drugs, but not with transdermal nitroglycerol.

The exclusion criteria were:
   Unstable angina pectoris
   Valvular defects
   Risk of myocardial infarction
   Myocardial infarction within the last 3 months
   Arrhythmia of clinical relevance
   Uncompensated cardiac failure
   Lack of repolarization in V5 > 0.2 mV
   Hypersensitivity to nitrate compounds
   Allergy to adhesive patches

Of the 147 patients, seven were considered as dropouts in the final evaluation. The 72 males and 68 females who remained in the trial had an average age of 68 years, an average weight of 72 kg, and an average height of 168 cm; 19.3% were employed, 64.3% were retired and 16.4% were housewives.

Diagnoses

Evaluation of the severity of the coronary heart disease (according to the classification NYHA) gave the following results: 56% of the patients were diagnosed as grade II and 44% as grade III. In many cases, this diagnosis was established on the basis of several criteria, either by typical anamnesis (136 cases), ergometric test (53 cases), and/or coronary angiography (24 cases).

Concomitant Diseases

Concomitant diseases were diagnosed in 114 patients: hypertension (63 cases), diabetes mellitus (31 cases), congestive heart disease (30 cases), and other diseases (72 cases).

Materials and Methods

Study Design

Since the effect of pharmacological intervention on the quality of life can be determined only with outpatients, the study was designed as a multicenter trial with 22 general practitioners in Austria. In order to compare the traditional oral therapy with nitrate compounds and the transdermal therapy with Nitroderm TTS, patients were assigned at random to two groups: in one group, the patients changed to the transdermal form of therapy, while in the control group the patients continued with oral medication.

The trial was open and lasted 4 weeks. As shown in Fig. 19.1, the patients in the Nitroderm TTS group first were treated with a dose of 5 mg glyceryl trinitrate/day (Nitroderm TTS 5). They were then medically evaluated after 2 weeks and, when necessary, the dose was increased to 10 mg/day (Nitroderm TTS 10). The patients in the oral nitrate group continued with their previous treatment.

Two tests were performed to assess changes in quality of life: the first (baseline) at the beginning of the study, the second at the end of the 4 weeks of treatment. In parallel to the subjective rating of the quality of life by the patient, the physician's assessment of the efficacy and tolerance of the medication was evaluated by questionnaire.

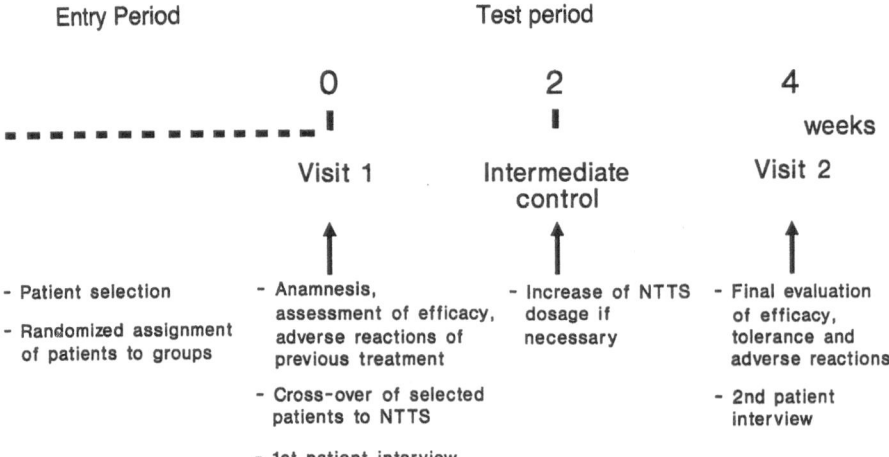

Fig. 19.1. Flow chart showing trial design

Quality of Life Assessment

For the purpose of this study the quality of life was assessed by two measures: a) the subjective assessment of well-being by the patient, b) the evaluation of conditions of life in general, i. e., the patient's ability to lead as normal a life as possible.

Two separate questionnaires, both designed for completion by the patient, were used to ascertain these two aspects of the patient's quality of life.

The first questionnaire, the so-called *adjective mood scale* according to von Zerssen [7], deals with the well-being of a patient. The scale is a list of 28 pairs of contrasting adjectives. The word pairs characterize opposite states of feeling, one describing a state of extreme well-being and the other its opposite. They assess in particular, mood (serious/light), motivation (lacking in drive/motivated), self-esteem (loved/unloved), and vitality (tired of life/enjoying life). Apart from the total score, the scale also yields separate subscores for particular aspects. Thus, groups of eight pairs are used to evaluate for example, "fatigue" or "depressive mood."

The second questionnaire deals with the *conditions of life* of the patient. It is a list of ten questions evaluating the patient's ability to walk, to climb stairs, to perform household duties, to participate in familial and social gatherings, and to enjoy recreational activities. Limitations are rated on a descending scale of five, from 0 to 4, 0 representing the optimal state. Like the criteria used for the evaluation of the patient's well-being, the various aspects of the "conditions of life" often overlap. For this reason, the overall score is a fairly reliable indicator. In addition, certain aspects can be analyzed separately (subscores): physical activity, social activities, and recreational activities.

Statistical Analysis

The statistical analysis was performed using standard methods (F-test, t-test, chi-square test).

Results

Randomization: Comparability of Nitroderm TTS and Referent Group

Table 19.1 shows the demographic and clinical data for the two groups of patients. There were no statistically significant differences between the two treatment groups with respect to demographic, clinical, or quality of life variables at the beginning of the treatment.

Control of Angina Pectoris Attacks

In the oral nitrate group, a slight decline in the number of attacks per week and in the consumption of fast-acting nitrates per week was observed after 4 weeks of treatment. However, neither of these decreases was statistically significant. In the Nitroderm

Table 19.1. Demographic and clinical data of the two treatment groups

Variable	Long-term oral nitrates ($n = 68$)	Nitroderm TTS ($n = 72$)
Demographic variables		
Sex (%)		
Males	54.4	48.6
Females	45.6	51.4
Age (years)	66	68
Weight (kg)	72	71
Height (cm)	169	167
Professional status (%)		
Employed	17.6	20.8
Retired	67.7	61.1
Housewife	14.7	18.1
Clinical variables		
Severity of CHD (%)		
NYHA II	61.8	50.0
NYHA III	38.2	50.0
Attacks/week (baseline)	4.8	5.6
Consumption of fast-acting nitrates/week	5.8	7.1
Concomitant diseases (%)	80.9	81.9

TTS group, on the other hand, the observed reductions in the number of attacks (approx. 2.93) and in the consumption of fast-acting nitrates (approx. 3.72) were both highly significant ($P < 0.01$).

Similar observations were made for the parameters "duration" and "intensity of the attacks." At the end of the 4-week trial, only one patient in the Nitroderm TTS group described the *duration* of the attacks as "increasing," whereas 71% classified them as "decreasing" and 14% as "identical." Of the patients treated with long-term nitrates, on the other hand, 7% classified the duration of their attacks as "increasing," only 24% as "decreasing," and 69% as "identical." This difference between the two treatment groups is highly significant ($P < 0.01$). As regards the *intensity* of attacks, at the end of the trial the percentage of patients with "decreasing" intensity in the Nitroderm TTS group was more than four times as high as in the group treated with oral nitrates. Again, this difference is highly significant ($P < 0.01$).

These results in favor of Nitroderm TTS were confirmed by the global assessment by the general practitioners on the efficacy of the two types of treatment. The therapy was estimated to be "good" for approximately 44% of the patients (see Fig. 19.2) in both groups; however, the score "very good," was attributed to 53% of the cases in the Nitroderm TTS group compared with only 12% in the oral nitrate group.

Adverse Reactions and the Incidence of Withdrawal from the Study

Adverse reactions were observed at comparable frequencies in both groups and followed the well-known pattern for nitro compounds. Only three patients had to

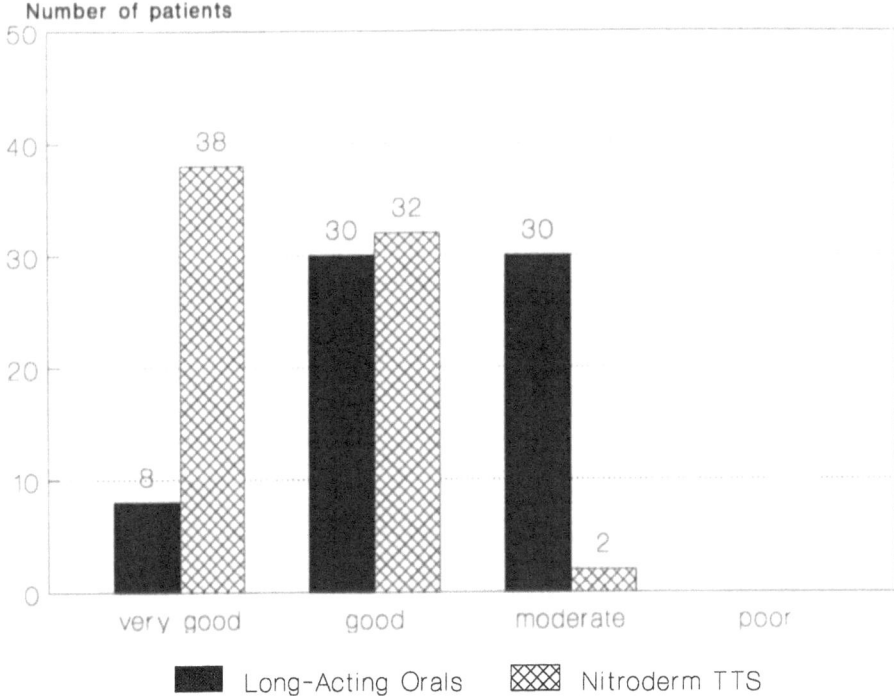

Fig. 19.2. Effectiveness of the treatment (investigator's judgment at the end of the trial)

withdraw from the trial due to adverse effects. Two of these cases were due to skin allergies to adhesive patches, and in the third case, treatment with Nitroderm TTS resulted in considerable deterioration in the patient's condition.

Quality of Life During Therapy

The results of the "quality of life" parameters are shown in Table 19.2. The table indicates mean values for the assessed parameters at the beginning and at the end of the trial for each treatment group, as well as the statistical significance of the observed changes. In addition, it shows the significance of the differences between the two treatment groups. It should be noted that a decrease in score indicates an improvement in quality of life.

Changes in Well-being Scores After 4 Weeks of Treatment

The analysis of the test results on the scale of well-being shows that Nitroderm TTS therapy led to a highly significant ($P < 0.01$) increase in the well-being of the patients. This was true not only for the total score of well-being but also for the partial scores "depressive mood" and "fatigue." In contrast, under long-acting oral nitrates, well-

Table 19.2. Changes in quality of life between the beginning and end of the trial (mean ± SD), for each treatment group

Parameter	Long-term nitrates		Nitroderm TTS		Comparison between groups

Well-being

Fatigue	↓ ▼	6.5 ± 3.9 / 7.3 ± 3.3	▲ ↑	7.7 ± 4.4 / 4.8 ± 3.7	
Depression	−	6.6 ± 3.2 / 6.6 ± 2.7	▲ ↑	7.1 ± 3.4 / 5.6 ± 2.4	
Total score	−	22.8 ± 11.2 / 24.3 ± 9.4	▲ ↑	25.9 ± 13.0 / 18.3 ± 9.2	**

Conditions of life

Physical activity	−	7.3 ± 3.4 / 7.4 ± 2.2	▲ ↑	7.9 ± 3.6 / 6.0 ± 2.1	
Social activity	−	4.1 ± 3.3 / 4.3 ± 1.7	▲ ↑	4.5 ± 3.1 / 3.2 ± 1.9	
Leisure activity	−	3.9 ± 2.1 / 3.9 ± 1.3	▲ ↑	4.0 ± 2.1 / 3.0 ± 1.6	
Total score	−	17.0 ± 8.6 / 17.4 ± 4.5	▲ ↑	18.0 ± 8.5 / 13.9 ± 4.7	**

↑ ▲, improvement; ↓ ▼, deterioration; −, no change; ↓ ↑ /*, $P < 0.05$; ▼ ▲/* *, $P < 0.01$

being of the patients either did not change (total score, depression) or deteriorated slightly (fatigue).

The results obtained with the scale of well-being allow an estimation of the extent to which a change in general occurred between "excellent well-being" (minimum score) and "extreme distress" (maximum score). They also permit comparison of the state of well-being of the subjects with that of the general population. According to the rating for the general population, test results corresponding to a Stanine value of six are considered "normal," seven indicate "moderate distress," and above eight "marked distress" prevails.

As can be seen in Fig. 19.3, the total scores of the two treatment groups, which differed slightly (though not significantly) before the start of treatment, correspond to a Stanine value of eight. This score is an indication of the substantial reduction in the well-being of the patients resulting from both the disease itself and the therapy. While

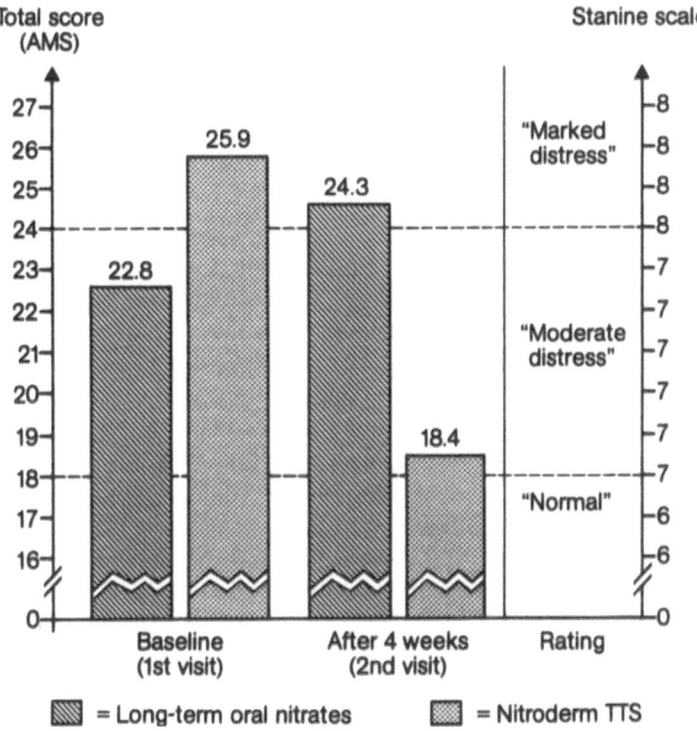

Fig. 19.3. Changes in well-being in the two treatment groups

patients treated with long-term oral nitrates remained in the range of "marked distress," those treated with Nitroderm TTS improved their state of well-being to close to "normal."

In conclusion, it can be stated that Nitroderm TTS therapy leads to a highly significant improvement in all analyzed aspects of well-being compared to baseline as well as to a reference therapy with oral nitrates.

Changes in the Conditions of Life After 4 Weeks of Treatment

The analysis of the results from the questionnaire on conditions of life (Table 19.2) clearly shows that the patients treated with long-acting oral nitrates maintained their restrictions on conditions of life throughout the trial. In contrast, the patients under Nitroderm TTS therapy experienced a marked improvement. This was true for the total score as well as for every subscore that was analyzed statistically. In the realm of physical mobility, this means, for example, that patients were able to climb stairs, walk longer distances, and were generally less restricted by their previous cautious attitude. Furthermore, their degree of participation in social events with family or friends or in public increased. Finally, Nitroderm TTS treatment allowed greater freedom in the pursuit of recreational activities and hobbies.

Thus, as far as conditions of life are concerned, it can be stated that the ability of Nitroderm TTS-treated patients to lead almost a normal life, only slightly or not at all impaired by their heart condition or adverse treatment reactions, is significantly greater than to that of patients on the oral form of therapy.

Discussion

Nowadays, before prescribing a drug therapy, a general practitioner has to evaluate carefully its effectiveness, efficacy, tolerance and cost as well as its potential for improving the quality of life of a patient. This latter aspect is becoming more and more important in view of the prevalence of chronic degenerative conditions. Whenever chronic treatment of a disease is required, the effects of medical intervention on the general well-being of the patient, his living conditions, and his life-style play a most important role in the selection of an appropriate therapy. Although general awareness of this situation is increasing, there is still a general lack of assessment of the quality of life in clinical studies.

In the present study, two types of treatment which are currently available for the prophylaxis of angina pectoris in Austria were compared with respect to their impact on aspects of quality of life. Over a 4-week study period, general well-being and conditions of life were rated by 68 patients with stable angina pectoris treated with oral nitrate compounds on the one hand, and by 72 patients undergoing transdermal therapy with Nitroderm TTS on the other.

A single daily application of Nitroderm TTS 5 or 10 led to a statistically highly significant improvement in all the parameters used in the assessment of quality of life. In contrast, patients treated with long-term oral nitrates experienced no changes in their state of well-being and their conditions of life.

The methodology chosen to assess changes in quality-of-life factors certainly did not include all possible aspects of such a complex concept as "quality of life"; i.e., important aspects such as sleeping behavior and sexual function were not evaluated. However, the questionnaires used in this study were chosen for particular reasons:
a) the usefulness of the adjective mood scale for assessing changes in well-being (its validity and sensitivity are statistically well documented);
b) the adjective mood scale is normalized, allowing comparison of the patient group with the general population [8].

Such a comparison is not possible with the questionnaire on the conditions of life, since this questionnaire was specifically developed for angina pectoris patients. The consistency and homogeneity of the results, however, confirmed our notion that both questionnaires are valid and sensitive instruments for evaluating the impact of angina therapy on quality of life. It should, however, be noted that the intellectual skills required to fill in the adjective mood scale (von Zerssen) represented a problem for some patients. The methodological design of the study may also give rise to questions concerning the validity of the results. The study design as an open trial with outpatients without a control group receiving a placebo was not aimed at scientifically establishing the efficacy and tolerance of a therapy. Our goal was to gather information on the conditions and results of treatment of angina pectoris patients in daily practice and under everyday conditions. In this respect, the present study shows that different means of treatment for angina pectoris are not only different in their clinical effectiveness but also in their effects on important aspects of well-being and everyday conditions of life of patients. By taking into account the potential effect of a medication on quality of life, the physician can choose a therapy which fulfills the wishes and expectations of patients to a much greater extent. Furthermore, by improving the quality of life, he can positively affect the physical and social life-style of a patient. This positive influence on the patient's well-being and daily life undoubtedly leads to improved compliance and thus adds to the long-term success of a therapy. It is our hope that more and more physicians will take into account the clinical, financial, and psychological significance of quality of life when looking for the therapy best suited to their patients.

References

1. Adverse reactions to bendrofluazide and propranolol for the treatment of mild hypertension: Report of Medical Research Council Working Party on Mild to Moderate Hypertension (1981) Lancet II:539–542
2. Jachuck SJ, Brierley H, Jachuck S, Willcox PM (1982) The effect of hypertensive drugs on the quality of life. J R Coll Gen Pract 32:103–105
3. Bullpit CJ (1982) Quality of life in hypertensive patients. In: Amery A, Fagard R, Lijnen P, Staessen J (eds) Hypertensive cardiovascular disease: pathophysiology and treatment. Martinus Nijhoff, The Hague, pp 929–948
4. Najmann JM, Levine S (1981) Evaluating the impact of medical care and technologies on the quality of life: a review and critique. Soc Sci Med 15F:105–107
5. Linden W (1984) Psychological perspectives of essential hypertension: etiology, maintenance and treatment. Karger, New York
6. Croog SH, Levine S, Testa MA et al. (1986) The effects of antihypertensive therapy on the quality of life. N Engl J Med 1657–1664
7. von Zerssen D (1985) Clinical self rating scales (CSRS) of the Munich Psychiatric Information System (Psychis München). In: Sartorius N, Ban TS (eds) Assessment of depression, Springer, Berlin Heidelberg New York
8. von Zerssen D (1976) Die Befindlichkeits-Skala: Manual. Beltz Test-Gesellschaft mbH, Weinheim

20. Physician and Patient Preference

B. HORISBERGER

Aim

This study represents a component of a transnational research program on the diffusion and application of Nitroderm TTS. The aim of the study was to analyze the discrepancy between the degree of acceptance of the Nitroderm patch and the differences in clinical effectiveness demonstrated for it. In contrast to other studies, however, the direct and indirect individual effects of the application of Nitroderm TTS by the doctor were to be determined in order to shed light on the discrepancy described. The different value judgments of the patients were to be compared with the doctors' arguments for applying the patch.

The design of the study (Fig. 20.1) used the interviewing technique, one of the classic tools of market research. Two hundred and fifty doctors, namely 107 GPs, 91 internists, and 52 physicians in hospitals, were interviewed personally on the basis of a structured questionnaire. In constrat to other market research studies on the subject of Nitroderm TTS, an attempt was made to obtain an objective overall picture through the use of rankings, point scales, and calculations [1].

Fig. 20.1. Design of the study

Therapeutic Objectives in the Treatment of Angina Pectoris

Method (Fig. 20.2)

In a first step, the objectives of the treatment of angina pectoris were ranked according to their level of importance on a 10-point scale. The interviewed doctors positioned their values between 10 ("extremely important") and 1 ("not important at all").

Step 1

Aim of therapy: Allocation of points

.

Importance of aim

Step 2

Aim of therapy:

Allocation of points for TNP and OLN

.

Degree of attainment

Step 3

Importance of aim: Degree of attainment

.

Fig. 20.2. Quantification of aims and attainment values   Index

In a second step, the doctors were asked to estimate to what extent the therapies with transdermal nitrate patch (TNP) and oral long-acting nitrates (OLNs) achieve the individual therapeutic targets set, i. e., the extent to which the therapeutic goals are met.

The evaluation of the degree of attainment was similarly carried out on the basis of a scale from 1 ("the therapeutic target is not reached") to 10 ("the therapeutic target is reached completely"). As in the first step, the average of the allotted points as a measure of the degree of attainment was calculated by the weighted arithmetic mean for each criterion.

In a third step, the values for the target importance were multiplied by the points for the degree of attainment. This resulted in a weighted degree of attainment (Index). If this calculating procedure is carried out for all the therapeutic targets, the individual values for each subtarget can be added together. The addition results in a point value as a yardstick for the global assessment of the selected or applied procedure [1].

Weighting of the Targets of Therapy

Corresponding to the importance of the targets in the treatment of angina pectoris, the doctors allotted a specific number of points to each of the individual therapeutic targets (Table 20.1). In the ranking sequence of the target criteria, doctors gave the prophylaxis of attacks of angina pectoris the highest priority:
– The reduction in attacks of angina pectoris (average of 9.4 points)
– The reduction of pain (9.0 points)
– The reduction in the risk of infarction (8.9 points)

In the second most important group, we find other target criteria relevant to the patients:

Table 20.1. Importance of different aims in the therapy for angina pectoris (basis: 250 doctors; scale: 10 = extraordinarily important; 1 = not important at all)

Aims	Importance of aims
Reduction of angina pectoris attacks	9.4
Reduction in pain	9.0
Reduction of risk of myocardial infarction	8.9
Provide feeling of therapeutic protection/ reduction of anxiety	8.2
Constant long-lasting effect	7.8
Comfort of medication/application form	7.7
Ensuring patient compliance	7.6
Increase of overal capability	7.5
Avoidance of hospitalization	7.3
Avoidance of absence from work	6.7
Reduction of need for acute medication	6.3
Avoidance of drug interactions	5.9
Avoidance of development of tolerance	5.9
Decrease of nondrug therapy (tests, cures, etc.)	5.5
Favorable cost-benefit relationship	5.4
Decrease of symptomatic medication (analgesics, tranquilizers, antidepressants, etc.)	5.4
Avoidance of side-effects	5.3
Decrease in concomitant antianginal medication (beta-blockers, calcium antagonists)	5.2

- Conveying the feeling of therapeutic protection (8.2)
- A patient-friendly medication (7.7)
- Increasing the exercise tolerance (7.3)
- Ensuring patient compliance (7.5)

   Most doctors ranked economic parameters last:
- Reducing the costs of nondrug therapy, tests, health cures, etc. (5.5)
- Favorable cost-benefit ratio (5.4)
- Reducing the amount of other medication used (5.4)
- Reducing additional antianginal therapy (5.2)

   Taken as a whole, it was found that the medical-therapeutic aspects were definitely of the greatest importance to Swiss doctors. Criteria which could be appreciated mainly by the patients, such as sustained prolonged action or patient-friendly medication, were also classified as being of considerable importance. Aspects relating to national or health insurance economics, such as a reduction in the use of other drugs and resources or a favorable cost-benefit ratio, were considered to be of less importance.

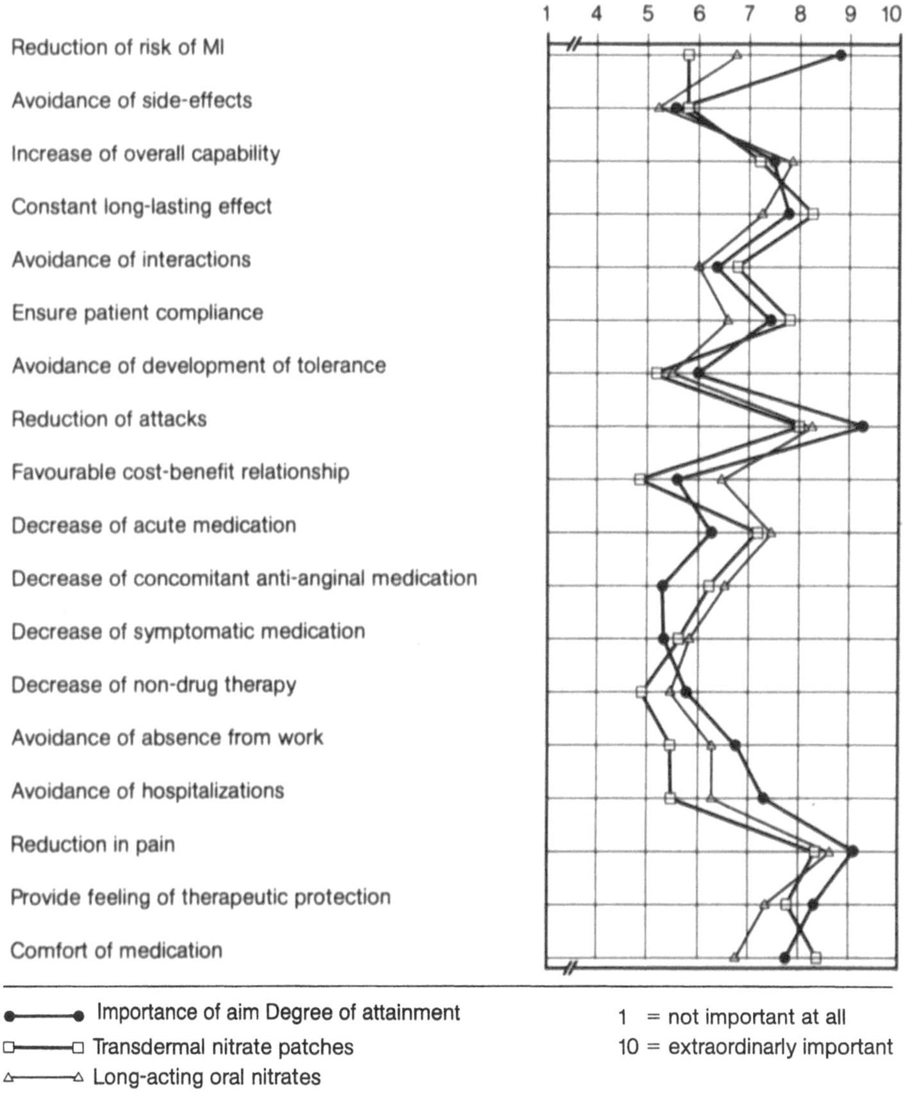
1 4 5 6 7 8 9 10
Reduction of risk of MI
Avoidance of side-effects
Increase of overall capability
Constant long-lasting effect
Avoidance of interactions
Ensure patient compliance
Avoidance of development of tolerance
Reduction of attacks
Favourable cost-benefit relationship
Decrease of acute medication
Decrease of concomitant anti-anginal medication
Decrease of symptomatic medication
Decrease of non-drug therapy
Avoidance of absence from work
Avoidance of hospitalizations
Reduction in pain
Provide feeling of therapeutic protection
Comfort of medication
●——● Importance of aim Degree of attainment
□——□ Transdermal nitrate patches
△——△ Long-acting oral nitrates
1 = not important at all
10 = extraordinarly important

Fig. 20.3 Profiles of target importance and degree of attainment

Degree of Target Attainment by OLNs and TNP

In the next step, the doctors allocated to the OLNs as well as to the TNPs[1] – again on the basis of a 10-point scale – the number of points which, in their opinions corresponded to the extent to which they attained the target (Fig. 20.3).

[1] At the time of the interview, apart from Nitroderm TTS, only one transdermal patch was represented on the market: Deponit (Searle), with a market share of less than 5%. The results for TNP are, therefore, virtually identical with what they would be for Nitroderm TTS

In a comparison of the profile of the therapeutic requirements with the corresponding profiles for the TNP and the OLNs, one is struck by the fact that rankings of 10 out of the 18 criteria of one of both groups of preparations are higher than the therapeutic targets, i.e., expectations were surpassed. For the three most important medical-therapeutic criteria, however, – reduction in attacks, reduction of pain, and reduced risk of infarction – the doctors found that the treatment with the two different products left something to be desired: they were not satisfied with the qualities of action of the two product groups. Both met the requirements with respect to reducing attacks and pain to an almost equal extent (moderately); however, the OLNs reached a relatively better result with respect to the reduced risk of infarction.

The three next most important criteria – assuring a feeling of therapeutic protection, sustained constant action, and patient-friendly medication – were better fulfilled by the TNP. The patch medication was also superior in ensuring patient compliance and in avoiding interactions and side-effects. Conversely, the OLNs were considered to increase the exercise tolerance slightly better than the patch. OLNs ranked better with regard to economic criteria. There was only a very slight difference in the ranking of the two product groups with respect to avoidance of the development of tolerance.

Comparison Between OLNs and TNP According to Groups of Therapeutic Targets

Out of 18 targets of therapy we now formed three different groups according to their orientations (Table 20.2):
– Seven medical-therapeutic criteria
– Four patient-related criteria
– Seven economically orientated criteria

Multiplying the target importance in each case by the degree of target attainment, we obtained index figures which combine the importance of both parameters. This multiplication process has no effect on the relationship between the numbers, it rather serves to bring out differences more clearly and allows the addition of the assessment figures.

The comparison between the average values for the three groups of therapeutic targets shows that – although the highest scores were allocated to three of the medical criteria – taken as a whole, medical criteria ranked slightly behind the average importance of the patient-related aspects. On the other hand, economic criteria were considered to be of distinctly subordinate importance.

While in medical-therapeutic criteria the degree of target attainment was found to be virtually equivalent, marked differences became apparent in patient-related and economic criteria. The degree of target attainment for the group of patient-related parameters was clearly in favor of the TNP. Conversely, the OLNs were shown to be superior – although somewhat less markedly – with respect to economic criteria.

To sum up, the assessment of the Swiss doctors – both in general and also from a medical-therapeutic point of view – describes the two groups of products as being equivalent. The oral form of administration is considered to be more economic and the transdermal form the better solution for the patients.

Table 20.2. Quantification of degrees of attainment

Criteria	Aims of therapy	Importance of Aim	Transdermal nitrate		Orale nitrate	
			Degree of attainment	Index	Degree of attainment	Index
Medical/ therapeutic criteria	Reduction of angina pectoris attacks	9.4	8.1	76.1	8.4	79.0
	Reduction in pain	9.0	8.3	74.7	8.5	76.5
	Reduction of risk of myocardial infarction	8.9	5.9	52.5	6.8	60.5
	Constant long-lasting effect	7.8	8.1	63.2	7.4	57.7
	Avoidance of interactions	5.9	6.1	36.0	5.8	34.2
	Avoidance of development of tolerance	5.9	5.1	30.1	5.3	31.3
	Avoidance of side-effects	5.3	5.9	31.3	5.0	26.5
Arithmetic mean/ subtotal		7.5	6.8	363.9	6.8	365.7
Patient criteria	Provide feeling of therapeutic protection/ reduction of anxiety	8.2	7.9	64.8	7.3	59.9
	Comfort of medication/ application form	7.7	8.2	63.2	6.8	52.4
	Increase of overall capability	7.6	7.3	55.5	7.7	58.5
	Ensuring patient compliance	7.5	7.9	59.3	6.4	48.0
Arithmetic mean/ subtotal		7.8	7.8	242.8	7.1	218.8
Economic criteria	Avoidance of hospitalization	7.3	5.6	40.9	6.3	46.0
	Avoidance absence from work	6.7	5.5	36.9	6.2	41.5
	Reducing the need for acute medication	6.3	7.1	44.7	7.3	46.0
	Decrease of nondrug therapy	5.5	4.9	27.0	5.4	29.7
	Favorable cost-benefit relationship	5.4	4.8	26.0	6.4	34.6
	Decrease of concomitant therapy	5.4	5.6	30.2	5.7	30.8
	Decrease of other therapy	5.2	6.3	32.8	6.3	32.8
Arithmetic mean/ subtotal		6.0	5.7	238.5	6.2	261.4
Arithmetic mean of total/total		6.9	6.6	845.2	6.6	845.9

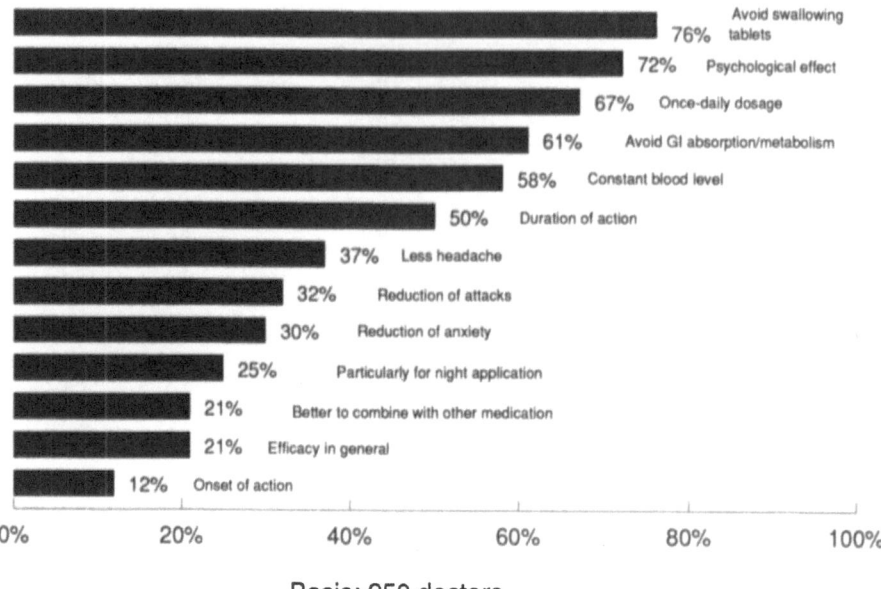

Basis: 250 doctors
(Multiple statements attached)

Fig. 20.4. Advantages of the transdermal nitrate therapy by comparison with the oral treatment with long-acting nitrates

In the context of the same study, the results found here were augmented by statements made by the physicians on patient reactions to Nitroderm and on the advantages and disadvantages of TNP compared with the OLNs.

Doctors' Acceptance

Advantages (Fig. 20.4)

Patient-related criteria again head the list of advantages:

- No need to swallow tablets (76%)
- Positive psychological effect (72%)
- Use once daily (67%)

Of the medical-therapeutic advantages ranked after the above, the following were considered to be of particular importance by the doctors:

- Avoidance of the gastrointestinal tract (61%)
- Constant blood level (58%)
- Duration of actin (50%)

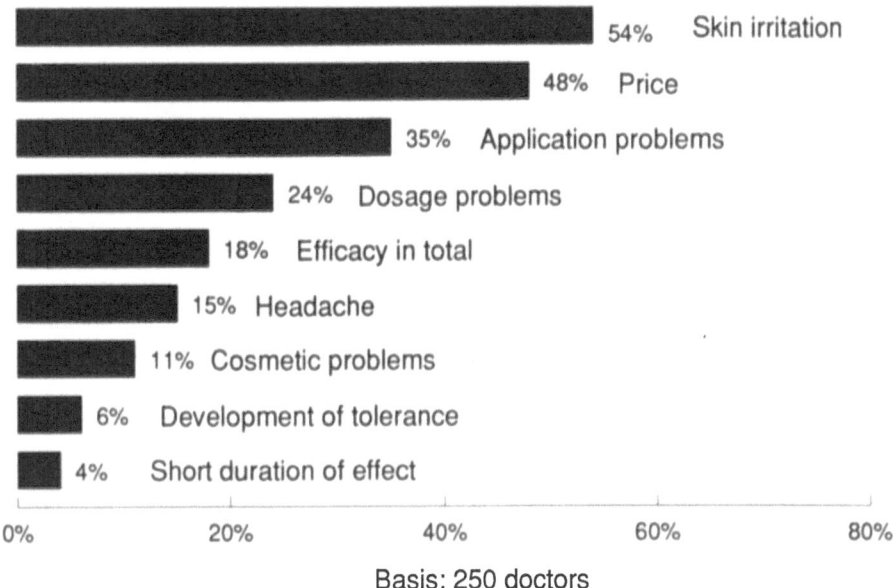

Basis: 250 doctors
(Multiple statements attached)

Fig. 20.5. Disadvantages of the transdermal nitrate therapy by comparison with the oral treatment with long-acting nitrates

The distaste for having to swallow tablets concerns those with a negative attitude to pharmaceutical preparations, but also those who have received polypharmaceutical treatments. The single daily dose concerns those who are unreliable in taking medication (compliance) and those for whom dosaging is difficult because of other diseases. Similarly, the avoidance of the gastrointestinal tract concerns those patients receiving multiple medication.

Disadvantages (Fig. 20.5)

The specific disadvantages of Nitroderm TTS compared with OLNs concentrated chiefly upon skin intolerance (54%) and the comparatively high price (48%). Technical aspects of the product, such as problems with application and dosage (35% and 24% respectively), also were notably criticized.

Only 18% of those questioned saw a relative disadvantage of Nitroderm TTS with respect to its efficacy; 21% saw a relative advantage. This points to the action of Nitroderm TTS being generally assessed as equivalent to that of the oral sustained release preparations. With respect to the problem of the development of tolerance under Nitroderm TTS – widely discussed in the scientific field – and its short duration of action, only 6% and 4%, respectively of the doctors saw a disadvantage by comparison with the reference medication.

If one compares the answers concerning the advantages and disadvantages, it is clear that considerably more doctors attributed more advantages to the TNP therapy than disadvantages, compared with the OLNs. Only three doctors said nothing about the advantages, whereas 14 doctors did not mention any disadvantages.

Patients' Acceptance (Fig. 20.6)

According to the statements made by the doctors, the reactions of the patients correspond largely to the description of Nitroderm TTS already given by the doctors in the rankings of the preparation according to the requirement criteria. Obviously, economic parameters are absent here, since these are not amenable to assessment by the patient.

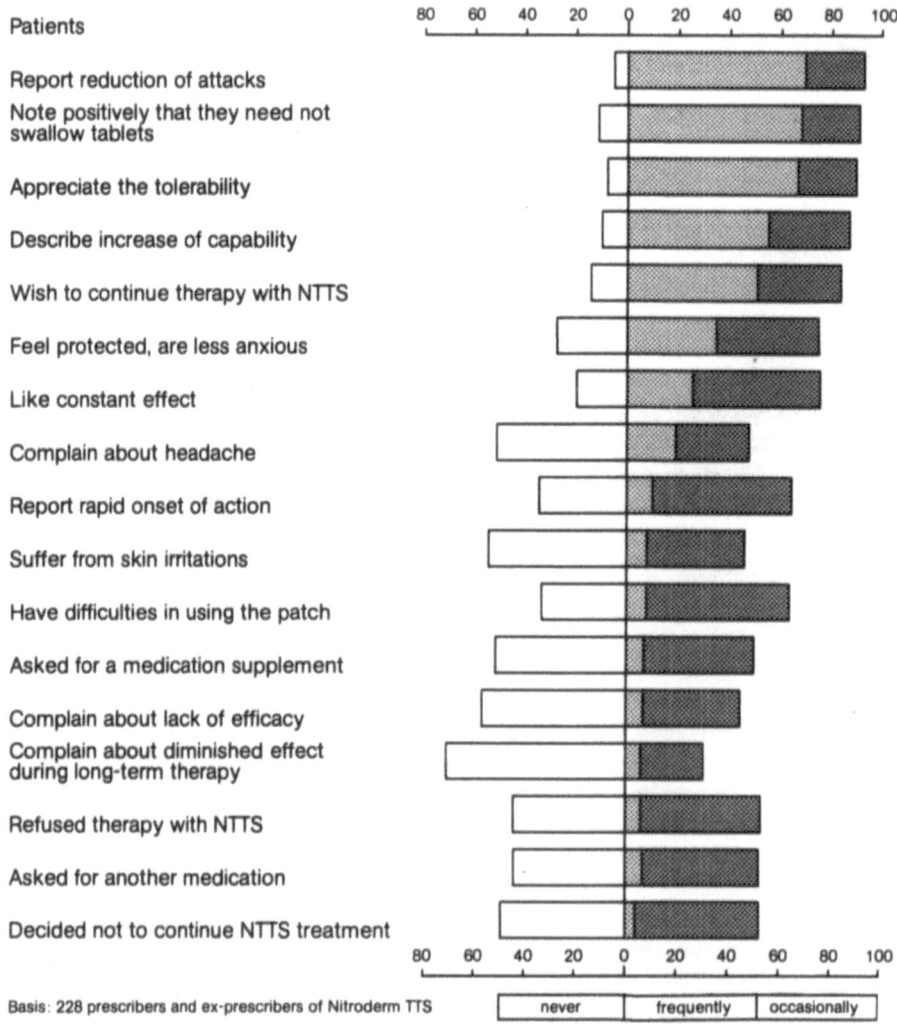

Basis: 228 prescribers and ex-prescribers of Nitroderm TTS

never | frequently | occasionally

Fig. 20.6. Patient response to treatment with Nitroderm TTS (*NTTS*)

Among the positive statements recorded by doctors, those most frequently cited were:

- Reported reduction of angina pectoris attacks
- No need to swallow tablets
- Improved exercise capacity
- Subjective feeling of protection, less anxiety
- Constant action appreciated

In only a few cases did the doctors find that patients expressly did not wish to continue with the Nitroderm TTS treatment. Inter alia, the fact that only about 2% of the doctors (frequently) observed the rejection or discontinuation of Nitroderm TTS therapy by the patient or that another form of medication was demanded, also shows that the rejection of the patch was contained within very narrow limits.

The doctors never heard or only occasionally heard negative statements by patients. Complaints were mostly about headaches, skin irritations, and problems of application. An inadequate or diminishing action hardly ever gave rise to criticism.

Summary

In essence, the results of this study demonstrate that the discrepancy mentioned in the introduction and investigated here can, in the case of Switzerland, be explained to a considerable extent by differing value judgments as to what constitutes rational and efficient drug therapy from a scientific, economic, and medical practice point of view. In the physician's day-to-day decision about the treatment of angina pectoris, the "in vitro" considerations of a clinical and economic nature are not accepted as binding preconditions.

The therapeutic decision taken by the doctor is found rather to be a synthesis of objective, scientific information, personal practical experience, and a case-specific assessment of the suitability of a treatment for the individual patient. In weighing the order of importance of science, practice, and patient, the doctor acts in accordance with a set of personal preferences which indicate his professionalism. These only partly agree – if at all – with the opinion in his environment. In the context of treatment for angina pectoris, the Swiss doctor reasons according to the categories of his patient. In the cases considered, the benefits to his patient and his psychological stabilization definitely represent the most important factors in the selection of objectives to be achieved through treatment.

There is no uniquely "correct" objective and no generally "correct" preference. In evaluative studies like these, practicing physicians, medical researchers, and patients may each have their own value system. The "right" objective is not a question of principle but of who the imagined patient is. His compliance with the doctor's decision may indeed depend on such external factors as the need to swallow pills once or twice a day, or the application of a patch. To this end, cost-benefit studies should try to take into account hard figures (e.g., those chargeable to insurance companies) and intangibles (those most important to the patient) [2].

References

1. Bapst L (1986) Die mehrdimensionale Kosten-Nutzen-Analyse als Evaluationsinstrument. In: Horisberger B, van Eimeren W (eds) Die Kosten-Nutzen-Analyse. Springer, Berlin Heidelberg New York, pp 37–39
2. Culyer AJ, Horisberger B (1983) Medical and economic evaluation: a postscript. In: Culyer AJ, Horisberger B (eds) Economic and medical evaluation of health care technologies. Springer, Berlin Heidelberg New York, pp 347–358

21. Patient Compliance in the Prophylaxis of Angina Pectoris

E. WEBER

Introduction

"Drugs don't work in patients who don't take them" – this is one way of stating the importance of patient compliance. Too few serious efforts have been made to assess this factor realistically in the course of clinical trials. Strictly speaking, it is not possible to rate the medical and pharmacological value of a drug correctly through a clinical investigation without knowing the extent of patient compliance. A socioeconomic evaluation and also the evaluation of quality of life are even more difficult to achieve without taking this factor into account. The methods developed to measure patient compliance are still rudimentary and serious attempts have not yet found their way into regular practice in clinical trials [2, 5].

Methodology and Design

The aim of the study presented here was to determine whether differences in patient compliance for a prescribed drug therapy could be detected between two patient groups receiving long-term prophylactic treatment for coronary heart disease: one on traditional long-term nitrates (LAO group) and the other on the newer transdermal therapeutic system: Nitroderm TTS. The study was carried out in the form of a prospective, open, controlled, randomized, interindividual multicenter trial.

Measurement of compliance was at the core of the study, which meant that it was extremely important to take experiences gained in previous studies into account if systematic errors in the quantitative determination of compliance were to be avoided. The parameter measured was the deviation by the patient from the prescribed dosage over a given period.

Three problems had to be considered:
1. How can the deviations from prescription be measured quantitatively?
2. When should the quantitative determination be performed and over what period?
3. Who should perform the measurements and how?

We chose the following approach:

1. The quantitative determination of patient compliance was arrived at by first counting and documenting the number and type of medications (individual dosages) in the hands of the patient. This count was repeated later, and the eventual deviation by the patient was determined by comparing the difference between the two counts with the drugs and dosages prescribed by the physician for this time interval.

2. A more difficult problem was to establish the exact time at which this quantitative determination should take place. In both treatment groups, the type of medication received for long-term prophylaxis of angina pectoris was new for every patient. For this reason, compliance might be expected to be better initially than later on in the course of a long-term treatment. This effect might lead to a systematic bias in the results. To avoid this effect without making the trial too long for practical purposes, a compromise was chosen, and the time point for the quantitative determination of compliance was set to begin 3 months after patient recruitment, with a 1-month observation period (Table 21.1).

3. Since a number of previous studies had shown that a reliable count of the tablets available to the patient cannot be achieved by asking him to bring them to visits, we performed the drug count and description at the patient's home in combination with two patient interviews. We decided to use a physician to perform this task, primarily in order to obtain further useful data of a descriptive nature, but also to give the patient the opportunity to discuss his experiences with a competent interviewer. The investigator's role was restricted to patient selection according to the inclusion and exclusion criteria and to the recording of anamnestic data.

This represents a novel study design; it is one of the few studies on compliance performed with outpatients so in Germany, and can, therefore, also be considered a pilot study. The exclusion and inclusion criteria were very strict, to ensure that multimorbid patients with at least one more chronic disease and complex therapeutic regimens (two to five concomitant medications) were recruited (Table 21.2). Furthermore, the inclusion and exclusion criteria made sure that the patient distribution was

Table 21.1. Study design

Selection of patients

1st Consultation
– Inclusion of the patient (acc. to inclusion criteria)
– Central randomization
– Changeover to the new therapy

Prestudy period (months 1–3)
– Getting the patient accustomed to new therapy
– Record or regimen changes

1st Patient interview (beginning of month 4)
– Initiation of investigation period (medication counting)
– Record of patient's attitude towards therapy and compliance, medication counting

2nd Patient interviews (end of month 4)
– Medication counting, calculation of differences between prescription and consumption of nitrates

Table 21.2. Inclusion and exclusion criteria

Inclusion criteria
- Exercise-induced stable angina pectoris treated with oral nitrates
- More than five angina pectoris attacks per week (untreated)
- Responsiveness to nitroglycerin during the attack
- Existence of other diseases to be treated with drugs
- Complex therapy regimen (between two and five different drugs, multiple administration)
- Patient is able to cooperate, to articulate, and is member of a GKV health insurance

Exclusion criteria
- Unstable angina pectoris
- Myocardial infarction in anamnesis (silent infarction allowed)
- Hospitalization during the last 3 months or longer
- Patch allergy
- Intolerance towards nitroglycerin preparations
- Previous treatment with nitrate patches

homogeneous with respect to sociodemographic and anamnestic data. It should be emphasized again that in both patient groups, long-term angina pectoris prophylaxis was changed when patients entered the trial. The patients were randomly allocated either to Nitroderm TTS or to another drug that was new to them.

General Data

Nine physicians from the Mannheim/Heidelberg region of the Federal Republic of Germany participated in the study (five specialists in internal medicine and four general practitioners). Thirty-four patients took part in the study, but nine of them discontinued treatment prematurely; among those, only one did so for reasons related to the nitrate medication. Thus, 25 cases could be evaluated.

Some sociodemographic data of the patients are given in Table 21.3.

Table 21.3. Sociodemographic data

Sex		Age		Marital status		Professional status	
M	F	50–65	Over 65	Single	Married/ partnership	Working	Not working
16	9	13	12	6	19	5	20

Figure 21.1 shows that the patients indeed presented the desired multimorbidity. On average, patients suffered from comorbid conditions besides coronary disease. The mean number of diseases per patient was 2.4. According to the NYHA classification for the severity of angina pectoris, 18 patients were in class III and seven in class IV.

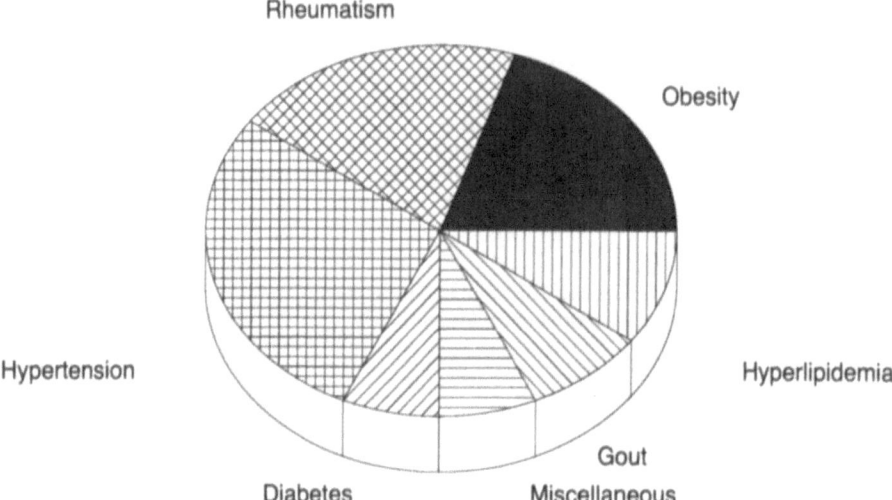

Fig. 21.1. Concomitant diseases (number of statements = 60)

Table 21.4. Medication for prophylaxis of angina pectoris

	Group A: Nitroderm TTS (n = 13) Patches	Group B: LAOs (n = 12) Tablets
Total prescriptions	320	605
Applied	297	490
Not applied	23	115

Compliance

The medication for attack prophylaxis is given in Table 21.4. It can be seen that even in the 4th month after the beginning of the study a higher proportion of the Nitroderm TTS patches were applied than orally administered nitrates, namely 92.8% versus 81.0%. This result is highly significant statistically ($P < 0.0001$). When compliance is rated according to three classes ($> 90\%$: good; $60\% - 89\%$: moderate; $< 60\%$: poor), the results presented in Fig. 21.2 are obtained. They show that in the Nitroderm TTS group compliance was always 60% or higher.

The results regarding the *concomitant medication* are listed in Table 21.5. The count was performed in the same way as for the medication for angina prophylaxis and again it can be seen that in the Nitroderm TTS group a markedly higher proportion of the prescribed drugs were taken: 90.7% against 73.3% in the LAO group. This difference is highly significant ($P < 0.0001$). Here too, more patients from the Nitroderm TTS group than from the LAO group fell into the two higher compliance classes.

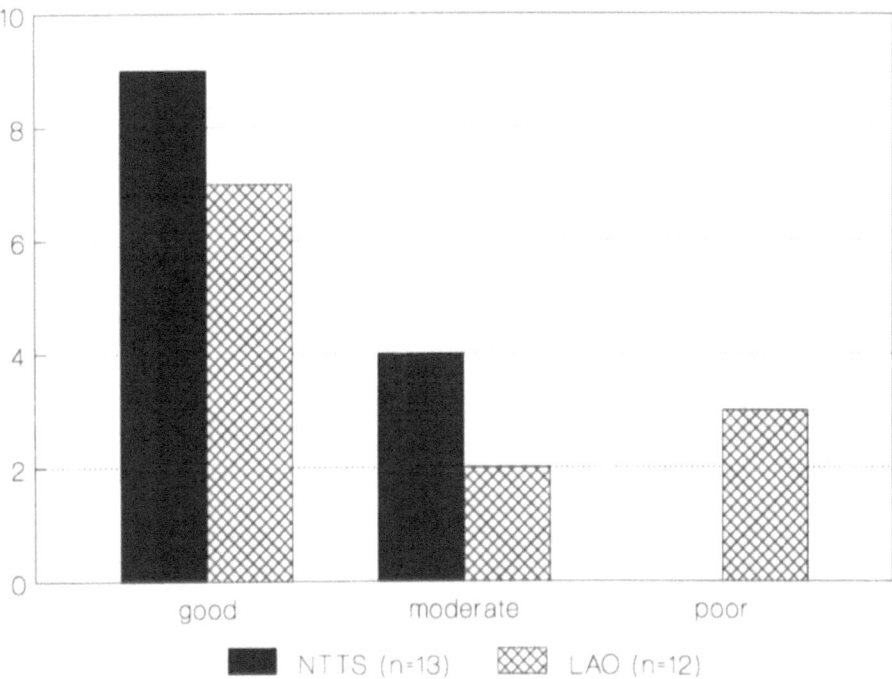

Fig. 21.2. Medication prophylaxis: evaluation by quality of compliance ($n = 25$)

Table 21.5. Concomitant medication[a]

	Group A: Nitroderm TTS ($n = 13$) Tablets	Group B: LAOs ($n = 12$) Tablets
Total prescription	2031	1245
Tablets taken	1843	912
Tablets not taken	188	333

[a] Sprays not counted

The results for the medication for angina prophylaxis and those for the concomitant medications are combined as total medication in Table 21.6. As can be expected from the results shown so far, the better compliance in the Nitroderm TTS group is again clearly visible. Only 9% of the prescribed drugs were not taken by the patients in the Nitroderm TTS group, whereas this number was 2.5 times higher in the LAO group, amounting to 24.2%.

A comparison between the objective results from medication counting and the evaluation of patient compliance by the physician as it emerges from a questionnaire confirms the assumption often found in the literature that physicians generally over-rate their patients' compliance [1, 3; for further references see 4].

Table 21.6. Total medication

	Group A: Nitroderm TTS (n = 13) Units	Group B: LAOs (n = 12) Units
Prescribed	2351	1850
Consumed	2140	1402
Not consumed	211	448

Undesirable Side-effects

The patients were asked at three times during the study whether they experienced undesirable side-effects which they attributed to their medicaments: at entry, and in the course of the first and the second interviews. At the same time, the undesirable side-effects were classified according to three degrees of severity. In both patient groups, severe and moderate side-effects diminished whereas the lighter side-effects increased (Figs. 21.3, 21.4).

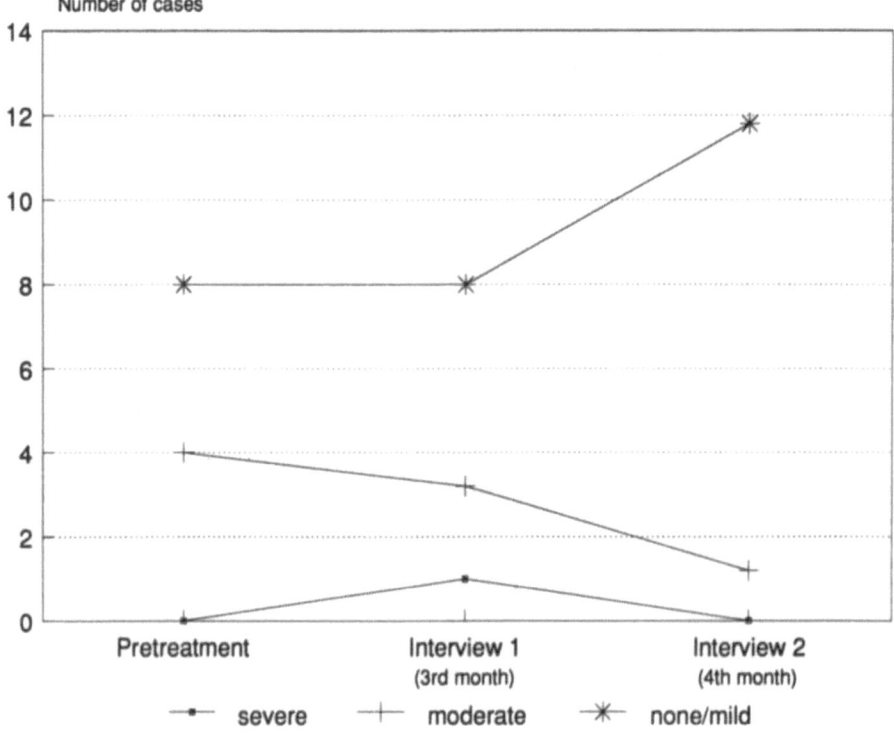

Fig. 21.3. Side-effects of LAO therapy

Discussion

The results relevant to the aims of the study are clear-cut and statistically highly significant. Both the medication for the prophylaxis of angina pectoris and concomitant medications were applied more reliably by the patients in the Nitroderm TTS group than by patients receiving an oral nitrate preparation for angina prophylaxis. Is there an explanation for this difference in patient behavior? To answer this question or to formulate a hypothesis, it is necessary to analyze the facts uncovered by the study which could have influenced its outcome. Apart from the quantitative parameters analyzed above, further data were collected for a descriptive evaluation. These were analyzed in order to see whether differences between the two groups appeared which could account for the better compliance in the Nitroderm TTS group. Statistical analyses showed a series of parameters to be distributed equally between the two groups which were not correlated with compliance. The only difference resided in the values for systolic and diastolic blood pressure, which were slightly lower in the Nitroderm TTS group than in the LAO group. This difference can hardly account for a difference in behavior in the two groups.

   An analysis of the overall medication regimens led to the further conclusion that the prescribed therapeutic programs were rather more complex in the Nitroderm TTS

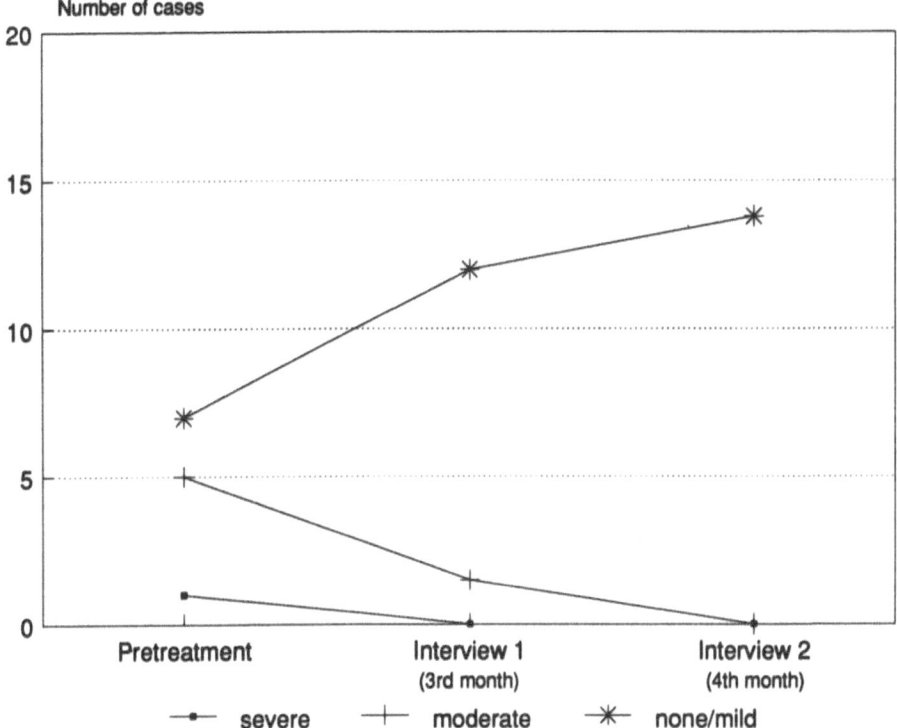

Fig. 21.4. Side-effects of Nitroderm TTS therapy

group compared to the LAO group. Both the total number of drugs prescribed and the frequency of daily applications were taken into account. It may be worth noting that in both groups compliance was rather better for more complex medication regimens.

Thus, these further investigations did not yield any indication as to why compliance in the Nitroderm TTS group should be better than in the comparison group. The only real and unambiguous difference is to be found in the form of application of the two drugs. This led us to the provisional conclusion that it is the prescription of Nitroderm TTS itself which leads to improved compliance, but this hypothesis will have to be substantiated by studies including a larger number of patients than the one presented here.

References

1. Davis M (1966) Variations in patients' compliance with doctors' orders: analysis in congruence between survey responses and results of empirical investigations. J Med Educ 41:1037–1048
2. Fletcher SW, Pappius EM, Harper SJ (1979) Measurement of medication compliance in a clinical setting. Comparison of the methods in patients prescribed digoxin. Arch Intern Med 139:635–638
3. Mushlin AI, Appel FA (1977) Diagnosing potential noncompliance. Physicians' ability in a behavioural dimension of medical care. Arch Intern Med 137:318–321
4. Weber E (1980) Patient compliance – a review of the recent literature. In: de Blecourt JJ (ed) Patient compliance. Hans Huber, Bern, pp 9–26
5. Weber E (1985) Compliance-Meßmethoden. In: Albus GP, Kaufmann W (eds) 3. Internationales Colloquium über Patienten-Compliance der Bayer AG. Verlag Medical Tribune, Wiesbaden, pp 48–65

22. Discussion of Part III

The contributions presented in Part III highlight many problems inherent in health services research, especially in the evaluation of the so-called subjective variables, or "soft data", as many would call them. According to one (invited) discussant, one should avoid the expression "subjective" as opposed to "objective" in reference to information. A more useful distinction, perhaps, would be "internal" and "external" information, in which case any subjective information from the patient (comfort, convenience, well-being, assurance, protection, etc.) would be termed "internal," and information from the treating physician (safety, reliability, efficacy–time relation, absence of side-effects, etc.) would be called "external."

Regarding these external variables, the arguments and the main interest are focused on four topics:
a) the quality of life,
b) the problem of measurement of subjective or social data,
c) the numerical expression of intangibles in the evaluation of alternative modalities of treatment, and
d) the planning of future studies.

Regarding the *quality of life* (QoL) argument, one should explicitly talk about "health-related" QoL with the intention of separating medical matters from environmental, political, aesthetic, and moral aspects. Although these may have an enormous impact upon QoL, they go beyond the assessment of a change in QoL following a specific treatment or comparing QoL as an outcome of treatment "A" as compared to treatment "B." Furthermore, that outcome, as a response to disease-oriented action or adverse effects, will vary from one patient to another and will be differently perceived. One should call it "patient-perceived" QoL. On the other hand, the perception of the treating physician about the patient's well-being and his (assumed) QoL will differ among doctors, and should, therefore, specifically be called "doctor-perceived" QoL judgment.

Regarding the "willingness to pay" argument as an indicator of the beneficial effect of a patch as compared with an (equally effective) oral medication, one must beware of cases where the answer is purely hypothetical. In cases where patients are asked about their willingness to pay, but the extra expenditure is covered by their health insurance, patients tend to overstate their readiness to accept the extra cost. On the other hand, when questions are phrased in such a way that the extra cost would be paid by the patient and not be reimbursable, the patients understate their willingness to pay. This change in opinion as a consequence of the phrasing of the question leads to the important problem of QoL assessment, namely methodology.

A good deal of the pessimism which seems to infect many of us when discussing QoL and matters related to it is due to *shortcomings in methodology*. The real question, surely, is the ease or the difficulty with which reliable[1] and valid[2] observations can be made. The value of any kind of observation in physics, physiology, or psychology depends mainly upon two factors: its relevance to the question that has been posed, and the accuracy (and precision) with which it can be (or has been) measured (see also Chap. 7).

Even many of those who know in principle how important it is to frame the questions carefully, seem to forget quite often that this includes adequately defining the terms in use: QoL especially can mean different things to government officials, sociologists, hospital administrators, doctors, patients and their families. In some evaluation studies it is not clear whether the investigation is mainly interested in facts or in opinions. Can one really claim to know about patients' preferences if one has only asked physicians for their opinions about the patients? Can one even define patients' preferences in terms of their willingness to pay for treatment in a social environment where treatment is free anyway? Does improved QoL improve or reduce compliance? Or, is improved compliance always rewarded by improved QoL?

The third important topic discussed was about *treatment and outcome*. How long a period of treatment is required to produce a change in QoL? Obviously, it will depend upon the nature of the illness and the treatment but, regardless of these, it seems reasonable to ask whether measurement after only 1 month can really provide evidence of a change in a patient's quality of *life!* Whatever the duration of treatment, how often within the period should QoL be measured? Ideally, it is the area under the curve that we should like to measure, but seldom can. However, is one measurement before and one after treatment enough? If they differ, is it because the method is unreliable, rather than sensitive to change? Of course, there are ways of separating these and other sources of variations, but they are seldom used. Should analysis be by "on treatment," or "intention to treat"? Probably the latter for QoL trials. Should treatment be blind or open? Open, surely, unless in everyday practice the patient would be blind as to the treatment he or she is receiving. But whatever choice is made of methods, sample size, duration of treatment, statistical analysis, indeed of every feature of trial design, the reasons for these should be described in sufficient detail to make informed discussion possible, not to speak of replication of the experiment.

---

[1] Reliability has been used interchangeably with repeatability, constancy, reproducibility, and precision. The conceptual notion is the extent to which the same measures give the same results on repeated applications. When a particular measure varies at two points in time, or when the same measure done by different observers gives different results, it is not reliable. A measure cannot be valid if it is not reliable. It can, of course, be reliable without being valid, just precise.

[2] Validity refers to truth. The more valid a measure, the more closely it resembles the true state of nature. Validity is closely related to accuracy. There are many different concepts and terms for validating instruments, all intended to help in deciding how well attributes not directly measurable are qualified. Face and content validity refer to procedures that assess whether a measure appears to measure what is supposed to be measured.

Source for 1 and 2: Balaban DJ, Goldfarb NI (1983) Medical evaluation of health care technologies. In: Culyer AJ, Horisberger B (eds) Evaluation of health care technologies. Springer, Berlin Heidelberg New York Tokyo, pp 28–29

The necessary size of a sample is seldom easy to determine, especially when there is little or no previous evidence about the magnitude of the treatment effect to be expected. But the problem is not solved by ignoring it altogether. Indeed, its existence is only one of several reasons for carrying out "debugging" or pilot studies, especially whenever new methods or procedures are to be introduced.

Incidentally, although dissatisfaction with existing methods (including the lack of translation into the language in which one is working) may be justified, no improvement is made by developing "special" questionnaires, or even test batteries, if these are not anchored as firmly as possible in some kind of validation. There is no space here to describe the range of methods of validation that may be considered. "Face" or "self-evident" validity is much the weakest of these and can almost always be improved upon, although not without some cost in resources. Translation of instruments validated in the original language requires validation afresh in the new. This they seldom get.

Regarding the *planning of studies* and the proper study design, some participants in the audience argue that we are often confronted with studies whose design is less than perfect. If findings are going to be generalized, we should aim at objective parameters and people will be demanding controlled, double-blind studies. Now, if one medication is a capsule and the other a patch it will be difficult to fulfill the double-blind requirement (one would need a four-branch model), but of course, a crossover examination would be possible (three or six months on pills and six months with patch and vice versa). One discussant mentions the placebo effect when the doctor puts something on the patient's skin. As this is very different to swallowing something, it is the discussant's suspicion that there is an important component to what this patch does to people. To him it is quite conceivable that the placebo effect could provide a core part of the value, some sort of therapy, over a long period of time with a sustained sense of "I'm doing something for myself and this is giving me protection, destroying my fears about the next attack, giving me a sense of control over the disease and I can take it off if I get a side-effect." Another author of a QoL study argues that there are, of course, some problems in the design of such studies, but objected to the assumption that the placebo effect was not tested. He mentions that if patients are recalled and reassessed after wearing a patch for 4 weeks, 4 months, or 1 year, there is no difference in the answer regarding QoL. He concluded that a placebo effect would fade with time and that such an effect was minimal or absent.

Nevertheless, when planning future studies, one should probably do more prospective studies. The advantage of cohort, longitudinal, or natural history studies would be to eliminate a lot of short-term effects. The true value of medication "A" as compared to the alternative "B" can be assessed as a computation of the probability of a switch from "A" to "B" and from "B" to "A," and the reasons for doing so can be recorded. Disadvantages of the prospective design are that it takes a long time to obtain results, the costs are high, and it is administratively difficult to maintain long-term contacts with patients and their doctors.

Part IV

*Panel Discussion on the Outlook for
Socioeconomic Evaluation of Drug Therapy*

23. Outlook for Socioeconomic Evaluation of Drug Therapy: Science's Point of View

F. Beske

The present discussion on health politics in the Federal Republic of Germany is, as in many other countries, above all centered around the concern over financing health care. It is apparent that in this discussion suggestions for concrete measures often fall victim to divergent interests and to a lack of knowledge about the effects of measures and proposals on the total health care system or on parts of it.

A prerequisite for a health care system which is both need-oriented and affordable in the long term is, in these times of decreasing resources, the guarantee of more efficient behavior in health care. To ensure this, research into specific areas is necessary, including methods of cost-effectiveness analysis.

Who are the users or the addressees of cost-effectiveness analyses? Firstly, without doubt, they are the decision-makers in the health care system. This includes not only politicians and administrators but also doctors, who take decisions on individual measures and proposals. Other users are public opinion leaders and, perhaps predominant until now, health economists.

There are indications that the awareness within the German medical profession of the need for economical drug prescription is increasing. Efficiency estimations carried out by doctors themselves play a decisive role in this, although these estimations are, in general, purely based on individual judgment.

According to the Drug Prescription Report from the Scientific Institute of the German Insurance Fund (Institut der Ortskrankenkassen), for 13 substances the cheaper generics accounted for more than 50%. Examples were amoxicillin, doxycycline, metoclopramide and vincamine. Yet the increase in the prescription of generics by no means indicates an uncritical bias towards cheap drugs. If a newly introduced drug has obvious advantages over preparations previously prescribed for the same indication, this can also lead to an increase in the amount of the drug prescribed, even if it does not entail any savings for the health insurance scheme. As an example of this I would like to mention the H_2-antihistaminics, which are used in ulcer treatment. The Drug Prescription Report mentioned above shows an expansion in the total volume of prescriptions of this substance group. This can only be attributed to an increase in the prescription of ranitidine preparations. The prescription of cimetidine has decreased. Ranitidine is roughly as expensive as cimetidine, but the risk assessment, especially concerning interaction and endocrinological disorders, gave a more favorable result for ranitidine; this change in prescription behavior is thus in accordance with the aim of improved efficiency although no direct expenditure cuts are involved.

Drug producers should observe these tendencies closely and interpret them as an opportunity: new, similarly priced or even more expensive preparations can be expected to be well prescribed if some advantage, for instance a better risk-benefit assessment, can be proved using cost-efficiency analyses.

There are, at least in the Federal Republic of Germany, now methods for increasing the transparency of the drug market, such as the Price Comparison List and the Transparency List, which are edited by the Federal Health Office and, in addition to price comparisons, contain pharmacological evaluations. It can be expected that these methods, as well as improved pharmacological training of doctors, will increase the awareness of doctors of the need for more efficient drug therapy. It should, however, be remembered that the above-mentioned changes in doctors' prescription behavior are not, as a rule, due to an explicit cost-effectiveness analysis, but rather to the doctor's own personal judgment.

The acceptance of cost-effectiveness analyses carried out by research institutes for decision-makers in the health care system depends on several factors. First of all, the analysis results must be accurate. This fundamental element is by no means easy to achieve, since even in perfect analysis processes the result is dependent upon the closeness to reality of often unspoken premises which are present in every cost-effectiveness analysis. Further, if a cost-effectiveness analysis is to be accepted by the decision-maker, then the decisions should be prepared using the analysis results; they should not be anticipated. This requires considering a wide spectrum of possible decisions and explicit definition as well as clear identification of all values included in the analysis. Where cost-effectiveness analyses are to be used for the preparation of decisions which must be negotiated between partners with divergent interests, the neutrality of the research institution is a further prerequisite for acceptability. Finally, necessary characteristics of a convincing cost-effectiveness analysis are that the methodology and the results are comprehensible, clearly laid out and easy to apply.

Cost-effectiveness analyses are plagued with a variety of problems. One recognized problem and yet one which can only be solved in individual cases is that of measuring and quantifying the significance of costs and benefits which are difficult to express in monetary terms. The term "quality of life" can be used as an example of the problem. In our societies and in modern health care systems it is not sufficient to ask, for example, how mortality in a certain system develops, whether life has been saved, and whether life expectancy has been increased; rather one must also ask whether quality of life has been coupled with these years of life gained.

In order to "measure" the quality of life, scales for describing the emotional and physical conditions are necessary for a wide range of symptoms, illnesses and signs of well-being. Such scales for measuring a person's state of health can only be comparatively judged, thus underlining the importance of exact definitions of decision alternatives. Here, we are in need of methods that are not too difficult to apply and that yield results which are not too difficult to understand.

Further problems arise in particular with cost-effectiveness analyses designed to serve as decision aids over a long period. Since any analysis, for obvious reasons, can only employ those data and those alternative lines of action available at the time of its conception, there is always the danger that the analyses will become outdated "overnight" due to medical or technical innovation. Adaptation of an already established cost-effectiveness analysis to new developments can only be achieved when as many parameters as possible are included in the structuring of the patient group being considered and in the description of the chosen courses of action.

In order to facilitate a cost-effectiveness analysis, it might be advantageous, at least for individual indication areas, to devise standards for the patient population as well as measurement and assessment regulations for cost and benefit positions. If such standards were generally accepted, this could lead to an even better acceptance of cost-effectiveness analyses by target groups. There is, however, the danger that a premature standardization might hinder further development instruments for cost-effectiveness analysis.

I shall now turn to a problem which arises when cost-effectiveness investigations are used to assess services which provide care for a whole patient group. The intensity of care, or seen from another point of view, the degree of usage of such services, plays a decisive role in the efficiency assessment. Instead of giving mean values for cost and benefit, it is necessary in the end to know cost and benefit curves dependent upon the degree of usage. At least limits for costs and benefits are necessary which mark the line beyond which an increase in the amount of care provided reduces efficiency.

One must also bear in mind when considering cost-effectiveness analyses that with the introduction of new health services or drug therapies, an expansion of indications may take place which cannot be immediately assessed. This expansion of indications can lead to shifts in the original decision alternatives given and thus change the cost-effectiveness ratios of these measures. Two examples help illustrate this point: The H_2-antihistaminics, previously mentioned, have brought about a wide range of improvements for ulcer patients, from eliminating pain to preventing operations. It must, however, be inferred that these substances are used to a much greater extent than is medically necessary, for example in the case of gastritis, which can be conventionally controlled using antacids. Such an expansion of indications naturally reduces the cost-effectiveness ratio of this substance group. This is difficult to predict and to quantify. Another example is that of the kidney lithotriptor, which offers the possibility of preventing expensive, unpleasant and risky kidney stone operations. Apparently, the use of the lithotriptor is not limited to those cases which would previously have required surgery; it is also now being used in those cases where a diet or drugs would previously have been considered the appropriate therapy.

Finally I wish to refer to the report of the scientific commission in Germany that is advising on "concerted action in matters of health" (*Sachverständigenrat*). In this report, which has just been published, it is expressly stated that cost-effectiveness studies are necessary in order to judge the economy of prescriptions. It is to be expected that this report will have some importance in the restructuring of our health care services.

In summary, a string of hurdles must be overcome in order to implement cost-effectiveness analyses. However, without solid cost-effectiveness analyses, no health care system is conceivable in which both medical development based on innovations as well as the interests of the individual and society are taken into account. This will be even more important in the future, when in any society many spheres of life, including health, will demand resources that are scarce and when, in the health field, different services and modes of practice, and in the drug field different drugs, will be in competition with one another.

24. Outlook for Socioeconomic Evaluation of Drug Therapy: Society's Point of View

L. KAPRIO

I shall mention some of the international problems regarding the rational use of drugs by reference to the Nairobi Conference and its follow-up. The international society, as seen by WHO, has suffered a long-standing crisis of confidence related to drug policy. The 1985 Nairobi Conference on the Rational Use of Drugs hopefully solved that problem for the time being. There is now a relatively reasonable consensus of opinion among governments, industry, consumer groups, prescriber groups, and also developing countries.

The priority for developing countries is to use the Essential Drug Program as well as strengthening further their drug control agencies. The WHO-revised drug strategy is starting to work well. The WHO research program for Tropical Diseases, the WHO Diarrheal Diseases program, and the Human Reproduction program continue to be high priorities. Of course, there is a new problem which is very important. A special program in AIDS started on 1 February 1987. The WHO DRUG INFORMATION contains an enlarged document for everyone concerned who would like to have recent information from WHO. However, I will return to the problem of socioeconomic and quality of life concerns as related to medicaments. In Europe as a Regional Director, I worked in cooperation with the drug control agencies. We have had so-called *Schlangenbad* Meetings, discussing drug utilization studies.

At present in Europe, between 0.5% and 1% of children die during the first year of life. The infant mortality is very low. Ageing is a more important factor in European public health, medical care, and drug utilization. By the age of 65, 80% of the population are still alive. Life expectancy for women in Iceland, Japan, and also areas in Europe is nearly 80 years. Half of those who live until 80 reach 85, even though these are the years of highest mortality. Ageing is a problem, and increasingly so. Women continue to smoke and will die increasingly of lung cancer. However, more and more people will reach 65. They are not necessarily healthy, as we all know from the discussions today and yesterday on angina pectoris. This is one of the future problems of quality of life and medicaments for the elderly. I would just like to mention that in Europe WHO did a study of health conditions, services available, and satisfaction of age groups between 65 and 69, 70 and 74, and 75 and 79. It was done in 14 countries, in different centers from Finland, Sweden and Belgium in Northern and Western Europe to Greece and Italy in the south. We even included Kuwait in this study. And one of the very peculiar observations was that the better the services, the more dissatisfaction there was. In Greece and Kuwait there is traditional family care of the elderly, who were very satisfied just that there was someone with them. In very sophisticated communities in Sweden and Finland, there was dissatisfaction, although every official service was available.

There are already "trade unions" for the elderly. One of the former Swedish Trade Union leaders has been very active in the pensioners' associations. The problems are known and improvements in quality of life are demanded. The elderly are now well educated and are becoming more and more demanding of services. They are not accepting things that society or doctors are trying to tell them. There is now a WHO program of Health Protection of the Elderly. Problems of medication and its overuse, as well as problems of compliance, have been analyzed.

European targets for Health for All by the year 2000 recommend preventive, holistic health policies. The medical care side is only one aspect. The medical profession has its role, but other health professionals such as nurses and teachers also can be active. Participation of the people is needed for health promotion. Health promotion is a new, growing activity of many industrialized WHO member states as well as of WHO itself. It has some "greenish" aspects in some countries, "feminist" aspects in others. There is a clear movement toward people being self-reliant and understanding their responsibility in the use of medicaments. Some of the self-medication problems are related to this self-responsibility and search for a better quality of life. Participants in this symposium have discussed one aspect of the problem of quality of life, but there are similar problems in the whole medical care area. Cancer patients recover or are maintained more than 15 years after diagnosis of cancer. There are other rehabilitation problems and then there is the maintenance of people who cannot be cured but who can enjoy a certain type of quality of life with the help of medical and social support. The nurses, for example, are even more interested in the quality of life than are some physicians and surgeons. Chronic diseases, impairment and disability, and terminal care all demand serious attention with regard not only to the safety and efficacy of drugs but also to their utilization pattern and influence on quality of life. I very warmly support your studies, and any studies of this nature.

I have been present at many meetings such as this one here in Wolfsberg. I would like to thank the organizers for allowing me to make my general observations. My major message is that there is an increasing need to pay attention to the use of drugs and a need to help people use drugs rationally. My second message is that we must communicate, keeping contacts open between all concerned, from industry to the medical professions, but finally, with the patients, the sick, and their families. Society will watch the economic aspects of care. We all have to watch and safeguard the human aspects.

25. Outlook for Socioeconomic Evaluation of Drug Therapy: The Company's Point of View

W. P. von Wartburg

I have been pleased with the exchange of ideas, viewpoints, and experiences which has taken place during the past 2 days at Wolfsberg. We are in a new area, with emerging methodologies, and I think we need to realize that we may not be exactly sure of how best to assess the situation. We need to be open to innovation in medical technologies and aware of the need for an overall assessment of the health care field. It would be foolish to say, for example, let's standardize everything according to double-blind crossover randomized trials and exclude cost-effectiveness or quality of life studies which evaluate the comparative drug treatments.

In the presentations given within this symposium, there have been some very important observations made regarding the interrelationship of the different aspects of cost-effectiveness. We have looked at specific countries and studies showing the concerns of patients, health care providers, and those who pay the costs. Dr. Kaprio placed the problem on a world scale by telling us about the worries of the World Health Organization. I, of course, approach the subject from the pharmaceutical company's perspective.

We have spoken of a morbidity which has to be dealt with by the whole health care system, not simply by one section of it. We are talking about prevention and treatment – both ambulatory treatment and hospital treatment. The pharmaceutical industry hopefully brings something to the health care table which will be of value in dealing with the morbidity of which we have spoken. Angina pectoris naturally needs to be treated on an individual basis, taking into consideration the doctor–patient relationship as well as the medications available. We are aiming at optimal disease management and this encompasses the authorities – public health, social security, or reimbursement authorities – and also the medical profession. Then there are the payers, who may be the patients, the social security institutions, or insurance companies. All want to have a say in the decision process leading to disease management – a decision which also involves the society's viewpoint.

More and more over the last few years patients have spoken out, wanting to be included in determining the process which affects them most directly. After all, they say, we are the consumers, the ones being treated.

This symposium has discussed quality of life studies and the approaches which affect the patient. The patient in the end does have the choice. Alternative forms of treatment must we weighed, both economically and in terms which may be less tangible but not necessarily less important. These are the quality of life considerations.

Professor Joyce spoke about the need to communicate – and the necessity of having information before we communicate. Here we come back to the pharmaceutical situation. The information we have generally supplied has been safety related and efficacy related; choices were made on those bases. Over the past couple of years the emphasis has shifted from the safety of drugs to the usefulness of drugs: their value for money.

To provide the information which is now most relevant, we need studies with different types of parameters, different types of country situations. We want to know how our drugs are behaving and this can be done in premarketing tests or in the analysis of drugs already on the market. We need to present our information and let the agency or person affected make an informed choice. It is my conviction that if you communicate enough information in a responsible way, you enable people to make good decisions, not just in medical terms but in overall terms.

The fact that every day one million patients are putting on a transdermal nitrate patch to manage their angina pectoris says a lot about the value of this new medication. However, more data need to be evaluated and communicated about the socioeconomic benefits of drug treatment, so that informed choices can be made by all parties involved. This was the aim of this symposium to which you have all contributed in a very open and stimulating way. Thank you very much and good-bye.

26. Editors' Note

We would like to start by commenting briefly on some points stressed by the first speaker on the panel (Beske, Kiel). Although the discussion of health politics in the Federal Republic of Germany and in other countries is centered on the concern for financing health care, the methods of socioeconomic evaluation of diagnostic and therapeutic procedures are not yet widely used in decision-making at the political level. Thus far, it seems that the phrase "value for money" is often understood in purely monetary terms: the cheaper the better. This has to do with the inherent difficulty in translating intangible values such as safety, comfort, general assurance, or well-being into "measures." The reluctance of the medical profession to adopt socioeconomic evaluation techniques and to accept them as complementary to epidemiological and clinical research stems to a certain extent from the difficulty in understanding the jargon in which such studies are quite often phrased. To overcome this hurdle, multidisciplinary discussions between doctors, health economists, and

health policymakers will be needed. We do not see any entrenched opposition of the medical profession against an evaluation of medical procedures, provided they are done scientifically and without undue generalization of the findings. We feel that the acceptance of cost-benefit analyses in health care will increase if we accept the limitations of the method. On the other hand, we must clearly go beyond cost–cost comparisons which are too often used in the cited "price comparison lists" or "transparency lists."

The proper application of a drug or any other medical modality depends upon a number of factors besides price, including
a) the correct diagnosis,
b) the appropriate indication,
c) the correct dosage,
d) the cooperation or compliance of the patient,
e) the doctor's interest in recording effect, progress, and side-effects.

Because of this multifactorial network of interdependencies in medical therapy, one should always aim at the result rather than limiting oneself to measurement of processes without evaluating the quality of outcome.

One major problem in evaluating progress in medicine lies in the fact that judgment is often needed *before* extensive experience is available. Such *ex ante* research is not without value if the result is taken as a hypothesis which has to be verified by forthcoming evaluation *ex post*. We are glad to hear that the scientific commission advising on "concerted action in matters of health" in Germany (*Sachverständigenrat*) has endorsed cost-effectiveness studies on the prescription of drugs. One hopes this authority is well aware that cost-effectiveness and cost-benefit analyses need to be constantly reconsidered and that evaluation is an iterative process in itself.

Society's point of view as pointed out by the next speaker (Kaprio, Geneva) focuses mainly on the quality-of-life aspect of health services and on health education. The tendency, at the international level, to support people's self-reliance and to promote self-responsibility in the struggle for better quality of life leads to the conclusion that the findings of evaluative studies should be translated into a message that can be understood by the community. To this end, close cooperation and communication are needed between all concerned, especially within the "magic triangle": doctor – patient – industry. Society and its politicians will continue to watch economic aspects of health services. Socioeconomic evaluation should contribute and help to supervise, promote, and safeguard the human and individual aspects of health care and to allow for choice. If one were not to allow choices to be made, of course, evaluation would not be needed.

The last speaker on the panel (von Wartburg, Basel) followed the recommendation of the previous discussant by adopting the holistic view of the health care system in trying to translate the problems of a pharmaceutical company into a social perspective. His starting point was a specific morbidity, coronary heart disease, and its leading symptom: angina pectoris. By asking about the value and the weight of this entity he took the bottom-up approach, starting with the individual patient and the doctor–patient relationship. Within this partnership, clinical decisions about optimal disease management are taken on the basis of information, knowledge, and experience. Later in the process of decision-making, many other parties become involved, representing public health, social security, and finally the interest of the payers of health services. In weighing all possibilities, he presents different possible alternatives for the treatment of angina pectoris and, at the transnational level, different situations in different countries. If the pharmaceutical industry is trying to improve the quality of life of patients suffering from angina pectoris, the open approach is to collect as much information as possible and to present the results for discussion.

Nowadays, roughly one million patients are putting a patch on their skin everyday to get relief from angina pectoris symptoms. There must obviously be good reasons for doing so, especially considering the fact that the patch is more expensive than the alternative. The application suggests that doctors and patients are convinced that this is good disease management, not just in medical terms but in individual patient-oriented terms as well. In this case it seems worthwhile to spend some time, effort, and money to find out the factors influencing such a decision. This is what socioeconomic evaluation is all about.

Of course, people at all levels in the health arena might oppose such studies by questioning their validity or their merits. But quite a number of decision-makers realize that good procedures can withstand the challenge of serious evaluation, and the pharmaceutical industry has, in the long run, a vital interest in supporting them. Evaluative studies depend crucially on appropriate design, conduct, and objectives. Equally important, though, is the communication between all interested parties: doctors, patients, economists, sociologists, politicians, and representatives of industry. However, before communication one needs information. Safety related, efficacy related, and cost-benefit related information in the broadest sense help to illustrate the forces weighing in the decisional process. Only with knowledge of these facts can one proceed to a reasoned assessment of the value of a medical modality. In this sense, studying the effects of applying a patch allows better judgment in a limited area. Good judgment in a broader context comes from factual knowledge of the components.

Part V

Envoi

27. Envoi

R. DINKEL, and B. HORISBERGER

Our purpose in organizing this seminar was threefold:
1. to evaluate the instrument of cost-benefit analysis from a broader perspective, namely not only from a methodological or technical point of view but also embedded in an environmental and social context;
2. to apply that methodology to the case of a new drug (Nitroderm TTS) which had been studied extensively and to the various social and economic conditions in six different countries, and to present data; and
3. to encourage an interdisciplinary dialogue between policy-makers, economists, sociologists, physicians, and drug manufacturers, in critically reviewing the data presented at the symposium.

As Joglekar and Paterson stated in a critical overview of cost-benefit analyses in the health-care sector: "The need for prudent use of scarce national resources in health care dictates that the social costs and benefits of alternative programs be measured carefully ... unfortunately, in practice, analysts have considerable leverage for choosing the objectives, assumptions, data, analytical methods, and interpretations that would yield desired conclusions" [1]. The validity of such investigations depends to a large extent on the researcher's readiness to choose not only reproducible and reliable but also "correct" data with corresponding face value. The impact of such investigations, on the other hand, depends to a large extent on the willingness of the partners in the game – manufacturers, doctors, patients, insurers, policy-makers – to accept socioeconomic arguments in the formulation of their policies and in their reactions.

The organizers wish to thank the participants for their criticisms and interventions in the course of the debate. Their contributions will serve to assist the methods of social and economic evaluation of drugs in the process of maturation. With this expectation in mind, the organizers hope that this volume will contribute to the improvement of evaluation studies on drugs yet to come.

References

1. Joglekar P, Paterson ML (1986) A typology of cost-benefit analyses in the health care sector. In: Horisberger B, van Eimeren W (eds) Die Kosten-Nutzen-Analyse. Methodik und Anwendung am Beispiel von Medikamenten. Springer, Berlin Heidelberg New York

28. Curriculum Vitae of Authors

BESKE, FRITZ
Prof. Dr. med.
Director of Institute for Health Systems Research, Kiel

1955 Master of Public Health at Ann Arbor, Michigan, USA
1958–1971 Medical Officer, from 1965 on Director in the State Department of Public Health /
State of Schleswig-Holstein
1961–1964 Medical Officer, European Office of the WHO, Copenhagen
1971–1981 Permanent Secretary of State of Schleswig-Holstein in the Ministry of Social Affairs
1972 Professor of Public Health and Social Medicine
Since 1976 Director of Institute for Health Systems Research in Kiel

BLOOM, BERNARD S.
Prof., Ph.D.
Research Associate Professor at the University of Pennsylvania

1961 B.A. Economics, Northeastern University College of Liberal Arts
1972 M.A. Economics, Northeastern University Graduate School of Arts and Sciences
1975 Ph.D., University of Pennsylvania, Graduate School of Arts and Sciences
1979–1981 Hahnemann Medical College and Hospital, Visiting Professor, Likoff Cardiovascular
Institute
1979–1982 Philadelphia Veterans Administration Medical Center, Chief of Health Services Research
and Development

DINKEL, ROLF
Dr. rer. pol.
Managing Director of HealthEcon Ltd., Basel, Switzerland

M.B.A. Economics, Ph.D. Economics
Studies at the Free University of Berlin, at the Universities of Munich and Wuerzburg. Director of
Dept. Organization and Planning, Glanzstoff Ltd., West-Germany (1969–1972)
Managing Consultant, Prognos Ltd., Basel (1973–1981)

GÜTHER, BERND
Dr. phil.
Project Manager at Infratest Health Research

Studies in Sociology, Psychology, and Economics

HANKIN, ROBERT
Administrator, Directorate-General for Internal Market and Industrial Affairs at the Commission
of the European Communities

B. A. Law
M. Phil. European Governmental Studies
1980–1983 Lecturer in Law, University of Birmingham
1983 Administration, Pharmaceutical Division, Commission of the E. C.

HORISBERGER, BRUNO
Dr. med.
Director of Interdisciplinary Research Centre for Public Health, St. Gallen, Switzerland

1952 graduated University Zurich
1957–1959 Research Fellow in Occupational Physiology at the Swiss Federal Institute of Technology
1959–1960 Research Fellow at the Chronic Disease Research Institute, Buffalo, N. Y.
1963–1979 Surgery and Head of Surgical Intensive Care Unit, Cantonal Hospital, St. Gallen
1971 Founder and Director of the Interdisciplinary Research Centre for Public Health
1980 Chief Medical Officer of the Canton of St. Gallen

JÖNSSON, BENGT E.
Prof., Ph. D.
Dept. of Health and Society, University of Linköping, Sweden

1969 M. A. Economics, Statistics and Political Science
1972 M. A. Economics
1976 Ph. D. Economics
1970–1982 Lecturer, Dept. of Economics, University of Lund
1979–1982 Director, IHE The Swedish Institute for Health Economics, Lund
Since 1982 Professor of Health Economics, Dept. of Health and Society, University of Linköping

KAPRIO, LEO A.
Prof., M. D.
World Health Organization

Special Adviser to the Director General, Headquaters, Geneva

1966–1984 Director of WHO Regional Office for Europe, Copenhagen

LAUPER, PETER E.
Dr. ès. sc. éc.
Head of Pharma-Economics at Ciba-Geigy Ltd., Basel

Lic. oec. at University of St. Gallen
Dr. ès. sc. éc. at University of Lausanne
Ciba-Geigy Argentina, Ciba-Geigy España, Ciba-Geigy do Brasil

LUCE, BRYAN R.
M. B. A., Ph. D.
Director, Medical Technology Assessment Program at Battelle Human Research Center,
Washington D. C., USA

NEUHAUSER, DUNCAN
Prof. of Epidemiology and Biostatistics, Dept. of Epidemiology and Biostatistics at Medical School,
Case Western Reserve University, Cleveland, Ohio, USA

1961 B.A. Harvard
1963 M.H.A. University of Michigan
1966 M.B.A. University of Chicago
1971 Ph.D. University of Chicago

OBERENDER, PETER O.
Prof. Dr. rer. pol.
Chair of Economics, University of Bayreuth

Studies of Economics and Social Sciences at the Universities of Erlangen-Nürnberg
(Diplom-Volkswirt), Munich, and Marburg (Ph.D.)
Director of the Research Institute for Health Economics and Social Law. University of Bayreuth
Chairman of the scientific group "Health Insurance" of the Robert Bosch Foundation

OSTER, GERRY
Ph.D.
Vice President and Senior Economist at Policy Analysis Inc., Brookline MA, USA

B.A. University of California, Santa Cruz
Ph.D. State University of New York, Stony Brook

READ, J. LEIGHTON
M.D.
Senior Research Associate for Health Policy at the New England Deaconess Hospital Harvard
Medical School

1973 B.A. Psychology/Biology, Rice University
1976 M.D., University of Texas
Internship, Internal Medicine, Medical Center, Duke University
Residency, Internal Medicine, Peter Bent Brigham Hospital
Fellowship, Center for the Analysis of Health Practices, Harvard Medical School

ROOS, BEAT A.
Prof. Dr. med.
Director, Federal Office of Public Health, Bern, Switzerland

Former Full-time Professor of Pathology, School of Medicine, University of Bern
Former Dean, School of Medicine, University of Bern

SCHNEIDER, HEINZ
Dr. sc. nat.
Project Manager at HealthEcon Ltd., Basel, Switzerland

1977 Ph.D., Biology at the Federal Institute of Technology (ETH), Zurich
1978–1979 Postdoctoral fellow, Departement of Anatomy, Laboratories for Cell Biology, University
of North Carolina at Chapel Hill

1979–1980 Research Associate Department of Anatomy, Laboratories for Cell Biology, University of North Carolina at Chapel Hill
Jan.-March 1981 Visiting Assistant Professor, Department of Anatomy, Laboratories for Cell Biology, University of North Carolina at Chapel Hill
1981–1982 Research Scientist Department of Pharmaceutical Research at F. Hoffmann-La Roche & Co., Ltd., Basel
1982–1986 Research Manager Xyrofin Ltd., Basel

TAYLOR, DAVID
Director of Public and Economic Affairs, ABPI London

1971 graduated in Sociology at the University of London
1976 Deputy Director of the Office of Health Economics
Since 1984 Director of Public and Economic Affairs, ABPI (Association of the British Pharmaceutical Industry), London
Since 1985 Chairman of the Family Practitioner Committee, responsible for Primary Care in the area of London
Since 1984 visiting senior Research Fellow at the Centre for Health Economics, University of York

TAYLOR, STANLEY H.
M.D.
Senior Lecturer in Cardiology at the University of Leeds

Undergraduate Education in Physiology and Medicine at the University of Birmingham
Since then various academic postgraduate posts in the Universities of London, Edinburgh, and Leeds

WALKER, STUART R.
Prof., Ph.D.
Professor within the Welsh School of Pharmacy, UWIST; Centre for Medicines Research at Carshalton, Surrey, England

Chelsea College, University of London: BSc II (1) Chemistry
St. Mary's Hospital Medical School, University of London: Ph.D.

VON WARTBURG, WALTER P.
Prof. Dr. jur.
Member of the Executive Management Committee of the Pharma Division and Head of the Department Pharma Policy and Economics of Ciba-Geigy Ltd., Basel

1963/64 Legal doctorate and bar exams at the University of Basel
1965/66 Legal and economic studies at Princeton University and Harvard Law School, Master of Laws degree
Professor on Public Health Policy at the Graduate School of Economics, Business and Public Administration, St. Gallen

WEBER, ELLEN
Prof. Dr. med.
Head of Department of Clinical Pharmacology, University of Heidelberg, Federal Republic of Germany

1955 M.D. University of Heidelberg
1965 Professor in Pharmacology and Toxicology, School of Medicine, University of Heidelberg
1969 Head of Dept. of Clinical Pharmacology at the Medical Hospital, University of Heidelberg
1974–1977 Dean, Clinical Medicine I, University of Heidelberg
1975–1979 Chairman of Clinical Pharmacology, German Pharmacological Association
Since 1977 Active member of Drug Commission of German Physicians, vice-member of Transparency
Commission and Drug Registration Commission A

WILENSKY, GAIL R.
Ph.D.
Vice President, Division of Health Affairs, Project HOPE, Center for Health Affairs

1964 A.B. University of Michigan, Psychology
1965 M.A. University of Michigan, Economics
1968 Ph.D. University of Michigan, Economics

ZWEIFEL, PETER
Prof. Dr.
Professor of Economic Policy, University of Zurich, Switzerland

1969 M.A. Economics, University of Zurich
1972 Ph.D. Economics, University of Zurich
1974–1975 Honorary Fellow to the Economic Dept., University of Wisconsin, Madison
1981–1982 Visiting Professor, University of the Armed Forces, Munich
1983 Assistant Professor for International Economics, University of Zurich
1984 Professor of Economic Policy, University of Zurich

Subject Index